A Sacred Place in the Enchanted Land

Galisteo Watershed
Shaded Relief

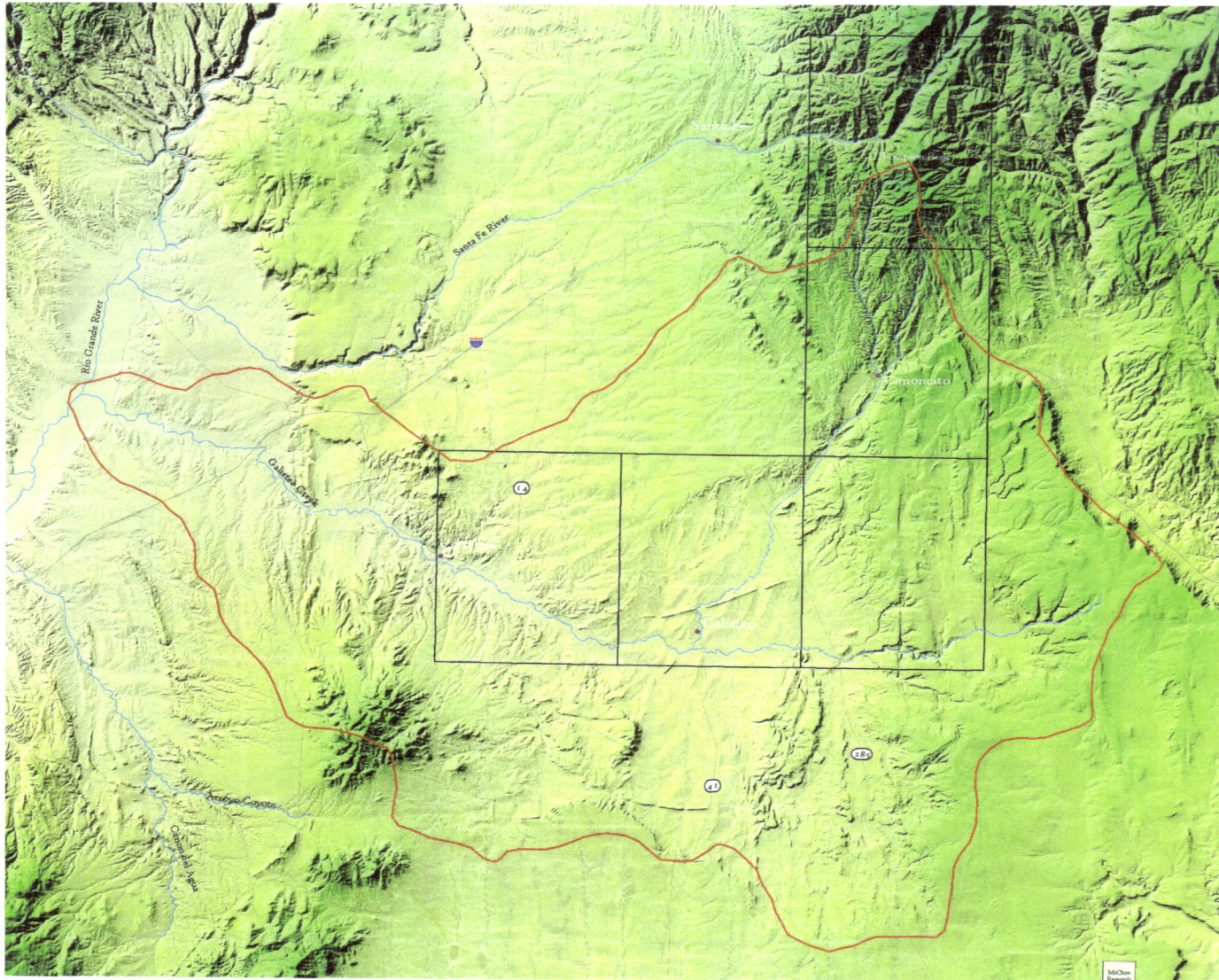

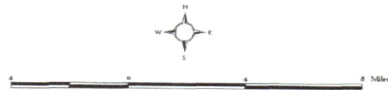

Santa Fe River

Rio Grande River

Galisteo Creek

Arroyo Canyon

Features

Towns

Approximate Watershed Boundary

Streams

Roads

Quad Names

| | | | McClure Reservoir |
| Picture Rock | Galisteo | Bull Canyon | Glorieta |

A Sacred Place in the Enchanted Land

Where its Heavenly Light Illuminates the Magic in its Dirt

PATRICK ALLEN MOHN

SUNSTONE PRESS

SANTA FE

On the cover: Cerro Pelon from Petroglyph Hill. Looking South from the top of Petroglyph Hill, one sees the mesa, Cerro Pelon, standing across the Galisteo Creek. From its summit on the left, or east side of Cerro Pelon in this photograph, a person can look across the long north south valley toward the Eastern Mesa Highlands. Very near to Cerro Pelon are several prehistoric Native American pueblo sites. In the near, right foreground, on a volcanic rock at the summit of Petroglyph Hill, are some prehistoric Native American Petroglyphs.

Sunstone books may be purchased for educational, business, or sales promotional use.
For information please write: Special Markets Department, Sunstone Press,
P.O. Box 2321, Santa Fe, New Mexico 87504-2321.

Design › R. Ahl
Printed on acid-free paper
∞

Library of Congress Cataloging-in-Publication Data

Names: Mohn, Patrick Allen, 1948- author.
Title: A sacred place in the enchanted land : where its heavenly light
  illuminates the magic in its dirt / by Patrick Allen Mohn.
Description: Santa Fe : Sunstone Press, [2021] | Summary: "The history,
  geology, ecology and beauty of the Galisteo Basin, a major geologic
  feature of northern New Mexico, are described, with many photographs by
  the author"-- Provided by publisher.
Identifiers: LCCN 2021039705 | ISBN 9781632933560 (paperback)
Subjects: LCSH: Geology--New Mexico--Galisteo Creek Basin. | Galisteo Creek
  Basin (N.M.)
Classification: LCC QE144.G35 M64 2021 | DDC 557.895--dc23
LC record available at https://lccn.loc.gov/2021039705

WWW.SUNSTONEPRESS.COM
SUNSTONE PRESS / POST OFFICE BOX 2321 / SANTA FE, NM 87504-2321 /USA
(505) 988-4418

I dedicate this book to my wife Dorothy, without whom it would never have been created.

CONTRIBUTORS

My thanks to those of you who have contributed to the creation of *A Sacred Place in the Enchanted Land*:

William Baxter, Homer Milford, Paul Secord, Ross Lockridge, Ann Murray

Francis Joan Mathien, PhD, archeologist

Sharon K Hull, PhD, geologist

Mostafa Fayek, PhD, geologist

Douglas Johnson, Jeff Salkeld who helped me get to this place, Jan-Willem Jansens, Yvonne Perea

Cathy Mora

Burnt Corn Pueblo, Conflict and Conflagration in the Galisteo Basin
by James Snead and Mark Allen

Todd Brown, Jerry West, Ted Perea

Sommer Betsworth and Mary Arrowsmith who provided my final proofreads and gave me great encouragement

Jason Mohn
I cannot thank you enough my son

and to

The Galisteo Basin Archeological Sites Protection Act

High Peaks of the Ortiz from Stagecoach Pass. Placer Mountain, which is the Ortiz Mountains highest peak, is to the left, and "The Western Peak" is to the right.

Green Grass and Wildflowers, Cerrillos Hills State Park.

Since 1970 I have lived within a geological feature known as the Galisteo Basin. Its natural beauty is stunning. It is a watershed defined by the drainages from surrounding mountains, mesas, hills, volcanic dykes, and ancient lava flows into the Galisteo Creek, which flash floods into a raging torrent during the summer monsoons. A heavenly light illuminates the Basin's broad, open vistas from its immense sky.

"Sunny Cerrillos," my friend Eric used to call our town. Sunshine is the most common weather. But there is every kind of weather, sometimes all on the same day, especially at springtime.

Once I saw a beautiful sunrise, with red in the clouds. The morning was warm and pleasant. Then the wind came up. A thunderstorm arrived. It got cold and the wind and the rain kept blowing. But, later, it got warm again. The wind stopped and the storm disappeared. Several of us climbed on top of the ancient wooden box car that used to be parked overlooking the San Marcos Arroyo, toward the west. We sat together and watched a marvelous sunset.

During the spring of the following year, I was walking my horse down a steep hill when it began to rain. As we continued down the hill the wind became strong and the rain turned to snow. I looked back at my horse and saw that the snow was caking heavily onto his windward side as lightning flashed very close overhead with a simultaneous loud clap of thunder.

The geology of the Basin includes a complex of Paleocene and Eocene sandstone, including the Diamond Tail and Galisteo Formations, and Eocene and Oligocene volcanoes. The volcanoes produced numerous volcanic dykes that radiate outward from them. These dykes protrude vertically through the sandstone.

In the upper Basin there are also some earlier sandstone formations, including the Mesa Verde Sandstone that forms the imposing mesa Cerro Pelon.

Within the Galisteo Formation there is a petrified forest. Large and small pieces of beautiful petrified wood can be found all around. I have seen places where each end of a log is protruding from the sandstone walls on either side of a small canyon or natural amphitheater, with the log's middle section lying in pieces where it had fallen to the sandstone floor in between. In one place, near the old Sweet Ranch, a very large petrified stone is the entire trunk of a tree.

And the sandstone of the Galisteo Formation comes in many colors and arrangements. The upper, eastern portion of the Galisteo Basin flows down from a broad, north to south expanse.

These drainages are dominated by Galisteo Creek, flowing from northeast to southwest, draining 10,554-foot Thompson Peak and other peaks in the Sangre de Cristo Mountains, and San Cristobal Creek, which originates upon a large highland named Rowe Mesa, much of which is covered with dense pinon and juniper forest, having very large trees. San Cristobal Creek flows down east to west, meandering through deep canyons, and then into a grassland valley before passing through the upper portion of the Eastern Mesa Highlands, a complex of tall, sandstone escarpments with projections of magnificent cliff faces of various colors. Several of the Basin's oldest Tanoan pueblos were built among these escarpments.

Galisteo Formation Sandstone with a Natural Window. This is one example of the great variety of color and arrangement found in the Galisteo Sandstone Formation. The individual who in 1859 had surveyed and then named the Garden of the Gods sandstone formation in Colorado, later surveyed this area of the Galisteo Sandstone Formation. He became so impressed by what he saw that he gave it its now official name: Little Garden of the Gods.

And then San Cristobal Creek enters the large valley before Cerro Pelon.

The two major drainages merge in this valley, south of the village of Galisteo and northeast of Cerro Pelon, and the two drainages then become Galisteo Creek.

From there the Galisteo Basin, flowing east to west, begins to narrow, most notably where it passes between two defining mountain ranges, both the remnants of ancient volcanos. These are the Ortiz Mountains to the south and the Cerrillos Hills to the north. Finally, after draining portions of the lava formations on La Bajada Mesa and Escarpment, the Galisteo Creek/River empties into the Rio Grande River.

Much of the Basin is pinon and juniper woodland transitioning to high desert savanna grassland at lower elevations and upon mesa tops, and all over the Cerrillos Hills. Drought tolerant, native species of grass grow together in various locations with varieties of cactus, yucca, sage, chamisa, pinon and juniper trees, and other plants.

Drought tolerant? The grasses that grow here appear light brown through the winter. If there are no spring rains, as often there are not, they will remain brown into the summer. But when significant rain does come, even if these rains have not come for years, the grass transforms the land to green: a beautiful, vibrant, shining green.

As you walk through this woodland/grassland, with its patches of clay, sand, and gravel soils, mudstones, volcanic formations, and its colorful sandstone outcrops, you see wildflowers.

Here are just some of those wildflowers that you see: Indian paintbrush, wild desert four o'clock, paper flower, baby white aster, desert purple verbena, sacred datura, desert baileya (desert marigold), bahia, prairie blue flax, deadly night shade (a naturalized plant), fragrant evening primrose, gourd blossom, varieties of penstemon, salsify or goats beard, narrow leaf globemallow, loco weed, Rocky Mountain beeplant, adonis blazing star, purple aster, snake weed, threadleaf groundsel, purple prairie clover, annual sunflower, narrow leaf yucca, broad leaf or Spanish bayonet yucca, New Mexico prickly pear cactus, the bright red blossoms of the claret cup cactus, pincushion cactus, and the translucent, magenta flowers of the walking stick cholla cactus. And some of the larger plants: chamisa, varieties of sage and saltbush, Apache plume, and desert holly.

These, along with many other native shrub, herb, cactus and plant varieties bloom in the dry, high desert climate; spring, summer, autumn, with many blooming during all of those seasons.

Cottonwood bosque, or forest, grows within the Galisteo River drainage along the creek, and in some of its tributaries, especially near springs. Alpine forests of ponderosa pine, Douglas fir, white fir, and aspen grow in both the Sangre de Cristo and the Ortiz Mountains.

The Galisteo Basin also has the unusual physical feature of being the synthesis, or coming together of four different eco-regions. This results in a greater diversity of terrain and plant life. And, as a transition zone, the Galisteo Basin is the migratory cross-path for numerous varieties of animal and bird species.

I have a memory of long ago walking down the San Marcos Arroyo from its springs. In the sky there was both thunderstorm and bright sunlight. I remember seeing a beautifully defined triple rainbow. I had already seen a number of double rainbows and knew that they were a very common sight in the Basin. I continued watching this rainbow as I walked. I thought, 'They even have triple rainbows here.'

I had no idea then how incredibly rare it is for anyone to see a triple rainbow anywhere on earth. Until fairly recently scientists theorized that triple rainbows did not exist upon the earth.

Desert Verbena, Ortiz Mountains. Just one example of the wildflowers that you see in the Galisteo Basin.

There are many treasures here. Whatever it is in life that you treasure, if you come here, you may find it. For me, a Promised Land.

But, as one might expect in such a place, the Galisteo Basin has a long history of mining. The first major gold rush within what is now the Western United States took place in the Ortiz Mountains beginning in 1821. Across the Basin, in the Cerrillos Hills, the first silver in the U.S. was extracted at around 1581 from a Native American Galina mine by Spanish explorers. Today in the Cerrillos Hills, the Cerrillos Mining District is an historic New Mexico Cultural Property. It embraces mines from the Spanish Colonial Period and from the Territorial Period, including the mining boom in the Cerrillos Hills beginning in 1879. But the most ancient mines of the Galisteo Basin are the turquoise mines in the Cerrillos Hills.

So, how ancient *is* the Galisteo Basin?

A large meteorite struck the earth. Dinosaurs were exterminated, except for birds, along with other animals, insects and plants, and in the seas fish, including species of sharks, and plesiosaurs, and corals, even the plankton. An estimated 75% or more of all species on earth disappeared during the Cretaceous-Paleogene extinction event 66 million years ago. At the same time, the lifting of the Laramide Orogeny had become high enough to drain the Western Interior Seaway off of North America. And so a new era brought a new place upon the earth.

Through the epochs of the Cenozoic Era the Galisteo Basin has formed and changed, draining the Sangre de Cristo Mountains of the Laramide Orogeny and other highlands. In the Basin, as everywhere on land, mammals, birds, and flowering plants found dominance while repopulating the earth's severely vacated ecosystems.

During the Paleocene and Eocene Epochs forests grew around streams and lakes in a warm climate. These would eventually form the sandstone, mudstone, and petrified wood of the Diamond Tail and Galisteo Sandstone Formations.

Horses began their evolution in North America. The two-foot-tall Eohippus would have browsed among the Basin forests, along with the seven-foot-tall flightless bird, Diatryma, and two-foot-tall Protylopus who looked, in miniature, much like the camels of today, but without a hump.

Plants would have included palms and magnolias.

The very first species of grass had already appeared by the time the climate began to cool toward the end of the Eocene Epoch. Forests were gradually replaced with savannas over millions of years, and modern species of grass evolved into being. Horses, camels, and other mammals adapted from eating a diet of leaves to eating grass. The abundance of available nutrients in grass and the proliferation of this family of flowering plants allowed mammals to grow larger, and evolve behaviors of migration. Thirty-four million and then again thirty million years ago the Cerrillos Volcano erupted. Twenty-nine million years ago the Ortiz Volcano erupted. While trying to imagine what the Ortiz Volcano once looked like, my friend, historian Bill Baxter, suggested to me, "Think of Mount Fuji!" These volcanos created great alluvial fans of detritus around their eruptive centers which buried the Galisteo Basin's southwesterly flowing streams. The movement of magma through the earth's lithosphere during these eruptions caused the Rio Grande Rift to begin its early phase. Fifteen million years ago, with the beginning of its second phase, the rift opened.

The density of the earth within the rift decreased as it widened, and the Sangre de Cristo Mountains, along the rift's eastern perimeter, floated upward as a block fault until, today, their ancient granite peaks are the highest in New Mexico.

On its west side the Rio Grande Rift spawned the Jemez Super Volcano, one of only six Super Volcanos in the world. Lava flows from an early Jemez eruption, perhaps twelve million years ago, now form part of the Galisteo Basin's northwestern boundary, La Bajada Mesa and Escarpment. Later eruptions, including 1.61 million years ago and 1.22 million years ago, sent volcanic ash as far away as Iowa. A large rock that blew out of Jemez was found in Kansas. The most recent eruption of the Jemez Mountains was a relatively small obsidian flow about 50,000 or 60,000 years ago.

The Los Cerrillos Mountain Range from an Ortiz Mesa, Sunset. From this location, looking north northwest, you have an interesting perspective for viewing the ancient volcanic structure of the Los Cerrillos Mountain Range, commonly called the Cerrillos Hills.

The south side of the Cerrillos Hills, which faces the camera's perspective, has, in particular, been eroded away by the Galisteo River which flows down, right to left, toward the Rio Grande River.

The major Cerrillos Hills Peak, Cerro Pelon, is not visible. From this perspective Cerro Pelon is located directly behind Grand Central Mountain on the right side of this image.

On the opposite side of the Los Cerrillos Range in this perspective, in its northeastern portion, alluvium from the Sangre de Cristo Mountains has flowed in, so that only the peaks of some mountains emerge out of the Ancha formation.

La Bajada Mesa and Escarpment from an Ortiz Mesa, sunset. This photograph, taken at the same time and location as the previous image of the Los Cerrillos Mountain Range, shows the large, very flat, lava topped La Bajada Mesa and Escarpment. Beyond La Bajada some of the southern portion of the Jemez Mountains can be seen along the horizon.

The Galisteo Basin began to tilt from its southerly course toward the west, in order to fill the Rio Grande Rift. As the Basin wound around it, from the north and east, "almost all of the body of the great Cerrillos Mountain went down the Galisteo and into the rift," according to Bill Baxter from a conversation with him, exposing deep layers of subterranean volcanics and the precious minerals of the volcano's interior. Then water, filling cracks in the rock, dissolved aluminum, iron, copper, and other minerals to form veins of turquoise.

Today, with most of the volcanic detritus that once surrounded the Cerrillos Volcano now gone, as well as a great deal of the Ortiz volcanic detritus (which is named the Espinosa Formation, and which comprise the Ortiz mesa highlands), the colorful and interesting layers of red mudstone and many colors of sandstone in the Galisteo Formation, and the thick, buff-colored layers of the Diamond Tail Formation, stand with their intricate complexity in the sunlight.

There is an extensive area of sandstone formations that were tilted upward, all those millions of years ago, by the force of magma, in a circular pattern around the Cerrillos Volcano's eruption perimeter. Sandstone layers stand vertical (even slightly past vertical in some places) at the inside of that perimeter, and then, gradually, at lesser angles farther back. Finally, the sandstone resumes its more horizontal layering after a few hundred yards. This circular pattern of the upward tilted sandstone is especially apparent from a distance. During the long spring times of the glacial ages the Galisteo Basin basked in the sparkling sunlight, draining the water of melting ice from some of North America's southernmost glaciers in the Sangre de Cristo Mountains, and pasturing the evolving megafauna. The Pleistocene Epoch was from 2,588,000 to 11,700 years ago.

Finally, man appeared, entering the Galisteo Basin while searching for the megafauna.

Many thousands of years went by. And then civilizations rose to power in Mesoamerica to the south, one after another. But the Basin remained thinly populated, with only a few people living in small pit house communities.

During the Developmental Period, early pueblos were constructed by the people living in these communities. Then, when Chaco Canyon and Mesa Verde were being abandoned, very large numbers of people settled into the Galisteo Basin: a civilization, a source for trade, many pueblos.

One site of Paleo-Indian presence was discovered in the Galisteo Basin dating to about 10,500 years ago. Mammoth and mastodon remains and camel tracks have been found nearby.

In the Sandia Mountains, just southwest of the Ortiz Mountains, Sandia Cave is one of the earliest known seasonal habitats for early man. Inside were found megafaunal remains of mastodons, mammoths, sloths, horses, along with mammal and bird species still living today. Also found were projectile points and scraps of baskets and woven yucca fiber sandals.

On the east slope of the Sandias, mastodon remains were discovered at 8,470 feet, the highest elevation ever recorded for this species.

The Ortiz Mountains from La Bajada Mesa. La Bajada Mesa and Escarpment is a broad, flat plain with some volcanic vents that once emitted its now dark lava surface. The mesa top's tall lava edges form steep cliff faces that have resisted erosion for the past twelve million years, preserving layers of Cerrillos volcanic detritus and Galisteo Formation Sandstone beneath the lava. In this photograph I am standing on the edge of the mesa, looking north to south across the western portion of the Galisteo Basin toward the location where the previous photograph was taken. The dark colored objects among the grass and plants in the foreground are lava rocks.

It is noteworthy that both of the archeological sites for which Paleo-Indian cultures were named are located in the eastern plains of New Mexico, less than two hundred miles away from the Galisteo Basin: Folsom to the northeast, Clovis to the southeast. The Galisteo Basin is a grassland corridor between New Mexico's eastern plains and the Rio Grande Valley.

The beautiful Clovis fluted projectile points and the other Clovis tools were designed for hunting and processing megafauna. Each large animal provided the Clovis people with a treasure trove of resources.

The Bering Land Bridge, or the continent named Beringia, which joined Northeastern Asia with Alaska, is believed by scientists to have existed during very cold portions of the last ice age, between 47,000 and 14,000 years ago, when the formations of ice lowered the earth's sea levels. During that time horses and camels, which evolved in North America, migrated across Beringia, into Asia and then Europe, while animals like mammoths came across in the other direction, into North America.

Between 25,000 and 15,000 years ago small, isolated groups of hunter gatherers followed these large herbivores into Alaska. Scientific evidence, including linguistic factors, blood type distributions, and molecular data such as DNA, link these people to eastern Siberian populations.

Between 18,500 and 15,500 years ago glaciers covering North America began to melt. Ice-free corridors developed along the Pacific coast and in valleys in the North American interior that allowed humans to migrate south from Alaska, on foot and by boat.

The recent discovery of a 14,000-year-old settlement in Canada has confirmed stories of the indigenous Heiltsuk Nation People. On Triquet Island, along the central coast of British Columbia, charcoal, tools, fish hooks, spears used to hunt marine mammals, and a drill used to start fires, support their stories, their 14,000-year-old collective memory of having lived there, and the idea that people first moved down an ice-free corridor along the Pacific west coast.

But people also migrated on foot down the corridor through west central Canada and into what is now the United States. The archeological site near Clovis, New Mexico was dated at approximately 13,500 years ago.

Upon their arrival into the Galisteo Basin, these people, from Siberia, Beringia, and then Alaska, would have seen an ecosystem that you and I have not seen: woolly mammoths, mastodons, Camelops the American camel, long horned bison, and herds of horses that were the size of modern day mustangs, giant ground sloths, the dire wolf (fearsome dog) the largest of all Canid species, the American cheetah, believed by scientists to be the reason why one of their prey species, the American Pronghorn antelope living today, evolved the ability to run 60 miles per hour; Smilodon, who is also known as the saber-toothed tiger (only distantly related to today's tigers), and the American lion who, at 25% larger than the modern day African lion, was one of the largest types of cat ever to have ever existed. Though technically not an antelope, that is, not related to antelopes of the Old World, the American pronghorn closely resembles them and fills a parallel ecological niche in North America. Today's pronghorn, a native species of the Galisteo Basin, is the only surviving genera member of twelve Antilocapridae species, along with species from three other now extinct genera of related animals. All of these species did exist, however, when the first humans entered North America.

Paleo-Indians were not numerous. Population densities were always quite low. Their lifestyle was arduous and dangerous, especially for children, pregnant women, and the elderly. There were large and fearsome predators. And, as with other nomadic cultures, people who were too injured or sick, or too old to continue the demands of the journey would likely have been, out of necessity, left behind.

Galisteo Formation Upward Tilted Sandstone. Like the previous image, Galisteo Formation Sandstone with a Natural Window, and located near to it, this sandstone had been tilted upward millions of years ago by the eruption of the Cerrillos Volcano.

But there were also celebrations. After migrating great distances, tracking the megafauna, then carefully and strategically stalking, perhaps at times unsuccessfully, in order to get close enough to kill the enormous mammals with hand thrown spears, there would have been reason to celebrate. The group would begin skinning and butchering and preserving everything. The enormous hides allowed for the construction of large, substantial shelters. (In Northern Russia and other locations in Northern Eurasia, large, igloo like shelters have been discovered that were made of large mammoth bones, some of these bones fresh at the time of construction, and some of them already old.)

One can imagine these people preparing for a celebration as they were all working together, processing the source of their livelihood. Perhaps at certain places and times of year several bands of people might meet for a celebration.

A cache of Clovis tools was discovered in Boulder, Colorado. Upon these tools was found the residue of horse tissue. The people who used them had been hunting and processing horses. They had cached the tools for the purpose of using them again later.

Scientists have identified a number of factors believed to have contributed to the Quaternary megafaunal extinction event in North America and Northern Eurasia at the end of the Pleistocene Epoch. They include climate change and hunting pressure by human beings.

The Clovis people, with their remarkable tool kit, were no doubt very successful hunter-gatherers. They may have contributed to megafaunal decline, and then extinction, in North America. Yet, human beings are known to have been hunting these megafaunal species, such as mastodons and mammoths, in Eurasia for more than 40,000 years. There is even evidence that Neanderthal man hunted these megafaunas in northern Eurasia. And there is no evidence that humans even hunted many of the megafaunal genera of the animals that disappeared.

Gradual climate change by itself appears to me even less likely to have been a primary cause for these extinctions. Many of these Quaternary megafaunal species had evolved in North America, while the American Mastodon had migrated to North America from Asia, possibly as long as 15 million years ago. During that time the megafauna would have endured periods of significant climate variability within the environment, including all of the glacial and interglacial periods during the Pleistocene Epoch.

There is, however, another scientific hypothesis that deserves to be mentioned.

Not far from the Galisteo Basin, at a famous Clovis archeological site: Murray Springs, in southeastern Arizona, the skeleton of a mammoth that was in the process of being butchered was found with a black mat over its bones. It appeared that someone was butchering this animal after it was killed at the time when this black mat covered it. Between the black mat and the bones there was evidence of an ejecta layer from an impact event.

Archeologists have identified this black mat at over 50 Clovis sites in North America. And they are markers of a major cataclysm.

The black mat consists of a layer of charcoal and soot that is lying directly upon a layer of microscopic diamonds, metallic microspherics, and other trace elements. This is the ejecta layer. Together, they indicate that a comet, or comet swarm, exploded in the atmosphere which then triggered widespread wildfires across North America.

It would have been a tough day for anyone, human or animal, in the Galisteo Basin.

The pattern of extra-terrestrial markers from different locations in North America, Greenland, and Western Europe indicates that the comet exploded over southern Canada, near the Great Lakes. No crater has been found. So, the theory is that approximately 12,900 years ago a comet about three miles in size exploded in the atmosphere.

Horizontal sandstone layers within the Galisteo Formation. This sandstone is located near the sandstone formations in the two previous images, but somewhat farther away from the Cerrillos Volcano's eruption perimeter.

The debris landed on the ice sheets, melting them, sending the water into the Atlantic Ocean. The influx of this vast quantity of water then changed oceanic circulations, which abruptly stopped the warming trend, and caused the Younger Dryas, an abnormally cold period from 12,900 years ago to 11,700 years ago; the coldest period, in fact, since the coldest part of the last ice age.

Many scientists are skeptical of the comet theory for lack of empirical evidence. But the Younger Dryas occurred suddenly, and 15 of 37 genera of Pleistocene mammal extinctions occurred 12,900 years ago. No doubt the theorized cataclysmic event itself, followed by continental wildfires, would have resulted in the instant death of people and animals, as well as the destruction of habitat, followed by 1,200 years of extreme cold.

These megafaunal species, along with others such as woolly rhinoceros, also went extinct in Northern Eurasia, relatively close to the believed impact site in Canada. Yet, modern species related to those that went extinct in North America and Northern Eurasia, such as elephants, camels, lions, cheetahs, and rhinoceros, did survive to live on in Central and Southern Eurasia, and especially in Africa, areas farthest away from the proposed impact site. Also, llamas, alpacas, and vicunas, relatives of the American Camel, live today in South America.

Nonetheless, the diversity in animal species numbers in the world today is very greatly diminished from it what had been prior to 12,900 years ago.

The extinction of the dinosaurs was well documented in the fossil record long before the theory was advanced by Luis and Walter Alvarez in 1980 that, based upon the discovery of iridium found within a thin layer of clay worldwide at the Cretaceous-Cenozoic geologic boundary, a six-to-eight-mile-wide asteroid crashed into the earth causing the Cretaceous-Tertiary mass extinction 65.5 million years ago. The 110-mile-long impact crater of that crash, located near the Yucatan Peninsula of Mexico, was not discovered until the 1990s.

On the other side of the earth from the Galisteo Basin there exists one of the most mysterious and enlightening archeological sites in the world.

Gobekli Tepe, located in the upper portion of the Fertile Crescent in what is now southeastern Turkey, a massive construction project even by modern standards, was built by stone age hunter gatherers beginning about 12,000 years ago. Encompassing 22 acres, the site includes multiple rings of large stone pillars up to sixteen feet tall and weighing six to ten tons each. These rings typically include two T-shaped pillars in the center, surrounded by a circle of slightly smaller pillars facing inward. Upon these pillars were carvings of animals, believed to be among the world's oldest pictograms. The pillars had been excavated from quarries several hundred yards away, and were then carried to the site and stood upright, the pillar bottoms having been carved to fit into notches that had been carved into the bedrock. It is estimated that up to 500 people would have been required to move, and then stand these pillars into place.

An abundance of stone tools was found at Gobekli Tepe, as well as the bones of game species such as gazelles. But the development of pottery, metallurgy, the wheel, agriculture, and animal husbandry had not yet occurred. So archeological concepts were upended. It had been assumed that the sociological changes to culture had *followed* the developmental changes. The construction of Gobekli Tepe, with stone tools by people who came together from small, isolated groups, indicates that the sociological changes had *preceded* the developmental changes.

The length of time between the constructions of Gobekli Tepe and that of Stonehenge in Britain is much greater than the length of time between the construction of Stonehenge and today.

Because no evidence of habitation such as cooking hearths or waste pits were found at Gobkli Tepe, archeologists have determined that small groups of hunter gatherers came together to create the oldest religious site yet discovered, and gathered there periodically for ceremonies. And it is thought that Gobekli Tepe was an observatory for monitoring the night skies.

Recently archeologists from the University of Edinburgh conducted a study of one of the stone obelisks. Pillar #43, known as the Vulture Stone, has carvings of vultures, scorpions, lions, snakes, boars, ducks, a wolf, and a headless man. One of the vultures has a circle carved above one of its wings.

It is known that hunter gatherers did not bury their dead, but conducted what is now known as the Tibetan Sky Burial: The bodies of the dead were placed upon the ground to be eaten by scavengers. It was believed that when vultures transported the flesh of the dead into the sky, they were taking the soul of that person into the heavens.

While studying the animal carvings made on the Vulture Stone the archeologists discovered that the creatures were actually astronomical symbols representing constellations and a comet. The constellations included Gemini, Pisces, Lupus, Ophiuchus, Sagittarius, Scorpio, and Libra.

The constellations of the night skies that we are familiar with today have their origins with stone age people from everywhere in world. Stone age people had been studying the night sky, the heavens, for an incalculable number of thousands of years, in great detail, observing the movements and patterns of movements of stars, planets, comets, the moon, meteors and meteorites, and the positions of the sunrise and sunset, with a very precise understanding of all of each of their various cycles, both through their own observations, and from the oral traditions of their ancestors.

The archeologists, using a computer program with their study, determined that the relative positions of constellations represented by animals on the Vulture Stone match to where they would have appeared in the night sky on a summer solstice about 12,950 years ago. This is also the exact time when ice sheet cores from Greenland indicate that the ice age known as the Younger Dryas began.

The recent finding of widespread platinum across the North American continent supports the theory of a giant comet strike. Scientists determined that this comet probably entered the inner solar system between twenty and thirty thousand years ago and would have been very visible as a prominent feature in the night sky.

It has been theorized that these stone age people who created Gobekli Tepe had a collective memory of the comet strike that had destroyed the abundant habitat of game animals and fields of wild wheat, barley and other seed plants which had, up to then, sustained these people so abundantly for at least the prior 10,000 years, and probably for a much longer period than that.

By the time agriculture began in that area, Gobekli Tepe had been long abandoned.

As their game species were disappearing during the Pleistocene Extinction, the focus of the Paleo-Indians turned upon the hunting of one particular species of bison, Bison antiquus. This transition strategy led to the Folsom Tradition of tool making beginning 11,500 years ago.

And then Bison antiquus became extinct. By about 8,000 years ago all of these species of megafauna, as well as other animal species, had essentially disappeared. The culture and technologies of the Paleo-Indian People also disappeared.

Mammoths became extinct about 11,500 years ago except for an isolated population that continued living on Wrangel Island in the Chukchi Sea. This population of mammoths finally disappeared 3,700 years ago.

In terms of the wildlife of the present day, I have a personal story to relate. One day Jeff told some of us that he had seen an eagle up in Eagle Canyon in the Ortiz Mountains. Although I knew that the Golden Eagle was supposed to be part of the Galisteo Basin's ecosystem, I had never seen one. So Skip, Jasper and I decided to attempt to see this eagle for ourselves. We climbed into Eagle Canyon during the morning, through the alpine forest, turning right, or south, where the canyon veers into the drainage of the canyon's origin: between Placer Mountain and the peak across the saddle from it, toward the west. And then we began climbing that mountain, the one we call the "Western Peak."

About two-thirds of the way up the mountain, we saw the golden eagle, soaring gracefully, from our left, just slightly below our eye level, gradually passing us, a calm, easy glide up the canyon. It had to have been aware of us. We could see in detail the feathers flaring out at its wing tips, and the individual feathers across the tops of its wings, back, and tail. We could see its head, its eyes. We were amazed at how large it was. It may have been a female. But as it neared the top of the canyon it turned toward its left, away from us, and passed through a flock of small birds. That event allowed us to see how really big the eagle was. It was an unforgettable experience.

The Sangre de Cristo Mountains from the Cerrillos Hills.

The Archaic culture developed from the Paleo-Indian traditions, and in the Southwestern United States it led, over a period of thousands of years, to the Ancestral Pueblo, or Anasazi Culture.

Many, many projectile points from the Archaic Culture have been discovered in the Galisteo Basin, atlatl dart point styles that reflect all parts of the Archaic time span, from 8,000 to 1,500 years ago.

The Archaic people utilized a greater diversity of plant and animal resources for food and technologies than their Paleo-Indian predecessors. Their nomadic patterns became more localized. It is said that the tool assemblage of the Archaic people differed so greatly from those of the Paleo-Indian people that no genuine continuity exists between them.

Perhaps there was a very difficult period for human beings, when Paleo-Indian strategies and tools were no longer useful, and Archaic Period strategies had not yet been developed.

The invention of the atlatl allowed for a smaller spear to be more precise, more powerful, and with a longer range than the large, hand thrown spears used by Paleo-Indian people. The Archaic people of the Galisteo Basin hunted jack rabbits, cotton tail rabbits, pack rats, turkey, mule deer, big horn sheep, pronghorn, and elk, among others. Predators in the Galisteo Basin would have included mountain lions, wolves, and grizzly bears, as well as black bears, bobcats, coyotes, and the elegantly beautiful, graceful and shy gray fox.

Rather than following megafauna over large distances, these people learned to exploit a broad spectrum of food sources, including a cultural focus on many plants. In this way they could minimize the risk of having a main source of food fail.

They migrated from place to place to gather plant resources as each matured, establishing a series of camps at specific collection points, returning to these locations year after year.

In the Galisteo Basin people gathered, among many others: pinon nuts, juniper berries, desert holly berries, and strawberries and raspberries in the mountains; rice grass seed for grain, prickly pear leaves and fruit, cholla cactus fruit, and yucca, both for its seeds and for yucca fiber to make basketry, clothing, and sandals. They wove nets and blankets utilizing yucca fibers, turkey feathers, and rabbit skins. Other plant sources of fiber were also utilized.

At first varieties of plant seeds were ground with specialized flat stones. And then manos and matates were developed, greatly increasing efficiency and digestibility.

The Archaic People of the Galisteo Basin also gathered medicinal herbs. This tradition of gathering and utilizing healing herbs and providing them to their people by healers, or Shamans, in conjunction with healing ceremonies, was practiced throughout North America by all of its hundreds of tribal groups. There were beliefs that the Great Spirit, or the Creator, gave these gifts to the people through Mother Earth, and that healing involved not just the herbs, but also living in harmony with the natural world. Gathering medicinal herbs and healing ceremonies continue to be practiced by Native People today, including the people of modern-day pueblos.

With basket making the people developed many important innovations. Food and other items could be gathered, transported, and stored. The art and technology of basketry became increasingly significant over time.

Wildflowers, Cerrillos Hills State Park.

One important food source for the Archaic people was Indian Rice Grass, a variety of bunch grass. It was so named because of its importance in Native American culture. Like rice has elsewhere in the world, this variety of bunchgrass provided a staple grain for the people of the Galisteo Basin, as well as for the people living in much of western North America at that time. The people created paddles with which they could paddle the grains out of the bunches of grass and into baskets.

Basketry and the creation of stone grinding tools allowed the Archaic People to transport and store more food and other items. Their development of more efficient grinding tools enabled greater nutritional sustenance from wild plant seeds.

A typical group of pre-agricultural hunter gatherers might consist of ten to fifty individuals. But these bands would join together with other bands from time to time, for ceremonies, feasts, and for mutual cooperation.

Over thousands of years Archaic people learned to adapt their foraging practices in order to tend wild plants in ways that encouraged those plants to produce more product with better reliably. The advent of this direct, more personal relationship between people and plants allowed the plants to change in response to human involvement with them.

The Archaic People began building shelters by digging shallow pits that were covered with a tipi shaped log frame. Over this they placed woven brush and plant fiber. Then the shelter was daubed with adobe clays.

Eventually, perhaps 5,000 years ago, these shallow basined lodges led to the development of Basketmaker pithouses.

Recent archeological research has pushed back the date of first maize agriculture in Mexico to 8,700 years ago. Maize, which has small cobs, is known to have originated from the domestication of a wild grass called Balsas teo sinte, which does not have cobs. It grows in the Balsas River Valley, where this research was conducted. By 7,000 years ago forests were being cut down there to create agricultural plots.

It may have been about 5,500 years ago that maize and squash were being grown in the American southwest. At some such time the Archaic Basketmaker people of the Galisteo Basin were planting maize corn near a water source. Afterward they would continue their annual migrations, then return at harvest time. They were efficient hunters and they focused upon that during winter months.

The Archaic people of the Galisteo Basin maintained their hunting and gathering lifestyle for a much longer period than Anasazis in many locations of the Southwest.

Eventually, their reliance on corn began to increase. Annual migrations became smaller. They developed sophisticated strategies for altering water sources to increase yield. These changes eventually led to the beginnings of pithouse communities.

The people began to line their baskets with clays that they found in alluvium soils. The accidental burning of these clays during the cooking of food may have led to the development of pottery.

Early pottery was made by lining a basket with clay, then firing it. This pottery was useful for boiling grains. Eventually the people learned to make pottery by coiling the clay, without the use of baskets. They learned to decorate their pottery with glaze made of minerals, and some researchers believe, also with plant dyes, especially from the Rocky Mountain Bee Plant, and other plants. It is thought that they learned to keep the plant dyes from fading off their pots by firing them in a low oxygen environment.

Quickly, the Archaic People became adept at making, decorating, firing and using pottery. They learned to dig clay from between sandstone layers. Unlike the alluvial clays, these clays from geological formations required a matrix of sand, finely ground rock, or ground up pottery shards, to be mixed into the clay and thus prevent the pots from cracking when they were fired.

The people learned to polish their pots, and then to coat them with a thin clay slip, creating various colors. The elegantly beautiful Black on White Ware, with its precise black drawings made with yucca fiber onto White Ware Pottery, was, and is today, very popular.

Development of pottery technology was contemporary with the people becoming less migratory. Baskets and other woven items were very helpful with migration. But pottery is easily broken. As the people gained insight into water management and became more reliant on farming, less upon hunting and gathering wild foods, pottery became the perfect fit.

This introduction of pottery, perhaps 1,500 years ago or more, began what we call the Anasazi Culture. I have a story to relate about one of the caves in the area from which these people obtained clay for their pottery.

One day, during the first years that I lived here, I was out hiking and decided to climb a very steep hill. This hill is actually the end of a couple of vertical layers of sandstone with a layer of clay between them.

I had been up there a number of times. Near the top was a cave in the layer of clay. My friend, Jeff, had told me that the cave was actually a prehistoric mine where Native Americans obtained clay for making pottery.

This time, when I looked in that cave, I saw and heard and smelled something back in the darkness: There were two creatures with long, bare necks that curved up and down in an opposing unison with one another as they hissed at me loudly. At first, I thought that they might be snakes, but I could make out their bodies down in the darkness, down at the bottom of their heads and necks.

And there was an obnoxious odor.

I tried to understand what I was seeing and hearing, but nothing made sense to me.

Standing on that precarious slope, I began to feel light headed and wondered if these were some kind of spirit creatures that had come up out of the earth.

And then I knew what I was seeing, although I have never seen anything like it before or since: These were two large vulture chicks, apparently on their nest at the back of the cave. They were trying to scare me away as they waited for their mother to return with food for them.

On top of La Bajada Mesa, at the highest part of the mesa top, is the La Cienega Pithouse Village. Built of basalt, or lava blocks, with unusually large room sizes, the community appears to have been both a domestic and ceremonial village. Dating of the pottery found there places the occupation during the Developmental Period, 100-1200 AD.

There are many petroglyph panels and trails along the mesa rim.

The Galisteo Basin is considered one of the most significant locations for archeological sites in the United States. In 2004 the United States Congress passed the Galisteo Basin Archeological Sites Protection Act specifically to safeguard them. They include at least ten major pueblo ruins, and thousands of outlying settlements.

San Cristobal Pueblo was built four or five stories high and is considered one of the largest pueblos in the Southwest.

By a count of ground floor rooms San Cristobal (1,645), San Lazaro (1,941), Galisteo (1,580) and San Marcos may be the four largest pueblos. The San Marcos Pueblo ruin, with its 2,000 ground floor rooms, was probably the largest prehistoric pueblo ever built in the United States. By comparison, the exquisitely designed stone architecture of Pueblo Bonito at Chaco Canyon contained 350 ground floor rooms.

Jeff Salkeld inside a natural cave in the Galisteo Basin. Photograph courtesy of Jeff.

One reason why the Galisteo Basin is not as well-known as other prehistoric Puebloan sites in the Southwest is that many of these sites are on private land.

In 1912 the great archeologist, Nels C. Nelson, having received permission from a local rancher, began excavating the site of the San Cristobal Pueblo ruins. While exploring midden deposits he developed his now well-known concept of *artificial stratigraphy*. He identified 17 room blocks and eleven plazas. He went on to excavate more than a dozen Galisteo Basin sites.

Many of the prehistoric mines in the Cerrillos Hills, where turquoise was extracted with stone tools, are well preserved because historic mining did not obliterate the ancient markings. Some of the most impressive of these mines are into Mount Chalchiuitl.

Chalchiuitl is Nahuatl, the language from the central Mexico highlands spoken by the Aztecs and Tlascalans. It can be translated as "precious green stone," conveying the power of the stone, not just its color.

Mount Chalchiuitl, a small mountain on the east side of the Cerrillos Hills, is, as a whole, the largest known prehistoric turquoise mine in North America. Native miners excavated thousands of tons of waste rock there with tools made of stone, leather, and wood. The use of sacred green stones has had a long history among Mesoamerican and Southwestern populations. By 1200 BC the Olmecs were constructing large platforms that they covered with green stones, then buried the floors. Jade was the most important stone for early Mesoamerican communities. Jade, mined in Guatemala, is not as abundant in Middle America as turquoise is in what is now the Southwestern United States. Turquoise was already in use in Mexico by as early as about 700 BC. As the populations grew in Northern Mexico and in the American Southwest, turquoise gradually became the predominantly used sacred green stone. (Cerrillos Hills turquoise comes in a variety of colors, including green, blue, and brown. Regardless of actual color, all of the Cerrillos Hills turquoise used by early Native Americans was considered part of their concept of sacred green stone.) No one knows for sure when turquoise was first mined in the Cerrillos Hills. Some archeologists have suggested that expeditions all the way from Teotihuacan in Central Mexico may have mined there as early as 350 AD. If that were true then one could infer that Cerrillos turquoise must already have had a reputation by then.

The earliest direct archeological evidence for turquoise mining in the Cerrillos Hills suggests a date prior to 900 AD.

Archeologists believe that an archeological site on the east side of the Cerrillos Hills, which includes an outdoor workshop and five structures, two of which were built with Chaco Canyon style masonry, and possibly two kivas, was a remote and seasonal outpost used by non-local people who mined the turquoise and then prepared it for transport. The absence of tools related to subsistence activities, such as manos, matates, and projectile points, indicate the site was used only for turquoise procurement, by people who brought their living necessities from somewhere else. Pottery sherds, also characteristic of Chaco culture, indicate that Mount Chalchiuitl was most intensively mined between 900 and 1140 AD, and, based on other kinds of sherds, again between 1300 and 1600 AD. The fluorescence of the Chaco Canyon Culture, which is about 100 miles west of the Cerrillos Hills mines, was between 900 and the 1140s AD, although turquoise was in use in Chaco as a trade item at least by about 750 AD. It has long been believed that much of the Cerrillos Hills turquoise was being transported to Chaco Canyon.

Originally a small pithouse farming community, the larger structures found in what is now Chaco Canyon National Historic Park in northwestern New Mexico appear to represent a very highly developed cultural and religious center that was primarily supported by the efforts of people living in thousands of other locations. Extensive amounts of turquoise artifacts in various stages of processing, and turquoise workshops, indicate that turquoise, and the creation of turquoise jewelry, and other exquisitely decorated turquoise items, was an important feature within the culture and an important trade item. Control of the turquoise trade, with its spiritual significance, may have helped Chaco to evolve. Mesoamerican trade items such as copper bells, macaws and macaw feathers, and chocolate were also present at Chaco.

Recently a doctoral student named Sharon K. Hull developed a technique that uses a secondary ion mass spectrometer (SIMS) to examine hydrogen and copper in turquoise. Archeologist Joan Mathien and Geologist Mostafa Fayek obtained funding for Hull's research.

With this technique Sharon Hull can examine a piece of turquoise and then determine its source. She has determined that turquoise found at Chaco Canyon did come from the Cerrillos Hills mines, as well as from other mines in the Southwest. Her research is ongoing.

Nomadic hunter gatherers lived in the Chaco Canyon area as early as 4,900 years ago. Ancient people began farming in Chaco Canyon about 200 AD and built small pithouses.

Already, by about 500 AD, turquoise beads and pendants were present at Chaco, and being used in ceremonies in great kivas. In 700 AD the population at Chaco may have been between just 100 and 200 people, a good sized pithouse community. During the 800s large construction projects were begun, massive stone structures unlike any that had been built before.

Wood for roof beams came from local pinon, juniper, and cottonwood trees. By the mid-900s the Chacoans were importing non local ponderosa, spruce and fir trees for construction. To this day Chaco Canyon's once naturally occurring pinon and juniper forest has not grown back.

During the 1000s construction of Pueblo Bonito and other Great Houses were completed, and the first use of chocolate outside of central Mexico appeared there in ceramic cylinders. By the beginning of the 1100s what has become known as The Chaco Phenomenon was well under way.

It has been discovered that the alignment of the architecture of the great houses at Chaco Canyon represents a sophisticated astronomical observatory. Its high desert location at 6,200 feet above sea level was ideal for this. Modern archaeoastronomers see from these alignments that the Chacoans had expert and clear knowledge of the cyclic and seasonal patterns of the sun, moon, planets, and stars.

Chaco is also famous for its unprecedented road system, built at a time and place where there were no such things as roads. They consisted of 180 miles of straight lanes more than 30 feet wide, precisely engineered upon the landscape, with berms and masonry walls in many locations.

The most famous North Road begins as a staircase in Chaco Canyon, cut into the rocks of the canyon wall near Pueblo Bonito and Chetro Ketl. Then, from the mesa top, the beginning of the North Road leads to Pueblo Alto and other locations, and on toward the Mesa Verde Civilization where there was abundant food and the manufacture of sophisticated ceremonial pottery. The North Road ends at the large and deep Kutz Canyon, where the remnants of a staircase then continues down the canyon's steep sides.

Two major roads went southwest to the forests of the Chuska Mountains. South roads were pointed toward present day Acoma and the prehistoric Guadalupe Community, as well as the forests of Mount Taylor, and toward present-day Zuni Pueblo. Beyond Zuni was the Mogollon Civilization, and beyond that, Central Mexico.

But the sophistication of these roads, at a time when there were no wheels or animals of burden, and their precise directional accuracy, along with the remnants of ceremonial activity and structures on the roadside believed to have been shrines, has led archeologists to conclude that Chaco's roads were constructed not just for practical purposes, but to link both Great Houses and smaller outliers across Chaco's large territory for ceremonial purposes. This concept is also supported by modern Pueblo beliefs about a North Road leading to *shipapu:* the place of their origin.

To this day new discoveries are taking place at Chaco. Recently, as of this writing, both male and female remains were located inside Pueblo Bonito, representing the burials of richly adorned citizens.

Then, in 1130, a fifty-year drought began, and by the 1140s the Ancestral Puebloan Culture at Chaco Canyon had collapsed. Other factors contributing to this collapse have been theorized. They include the development of a rich upper class and the dominance of a religious class that created social inequality.

There is no doubt that the Chaco phenomenon was driven, at least in part and possibly more than that, by trade with central Mexico. The Chacoan time frame corresponds with that of the Toltec Civilization, and turquoise was an item in great demand by the Toltecs. Toltec collapse was likely underway during the time of Chaco's collapse, though much uncertainty exists about the specifics of the Toltec Civilization. The Aztecs, who followed them, both idolized and imitated Toltec culture, and they destroyed much of what the Toltecs had created and built. The period of Toltec decline may have affected the decline of Chaco.

In 1275 another drought, combined with the need to support their much-expanded population, caused the abandonment of the large civilization on and near Mesa Verde in southwestern Colorado.

Again, other factors may have contributed to the Anasazi decline in the Four Corners region. Mystery surrounds the evidence of sudden abandonment, with food and belongings left in place, as if people were planning to return. Sunset Crater in northeastern Arizona erupted during this time period causing a long nuclear winter that may have devastated crop production for many years. And evidence of cannibalism in northeastern Arizona and southwestern Colorado has been discovered, a practice possibly imported from the Toltecs.

As people were migrating out of Chaco Canyon and Mesa Verde, new pueblos were being constructed in the Galisteo Basin.

Devil's Throne from across the Galisteo River. On this afternoon a rainstorm that had lasted for a couple of days was blown away by a strong west wind. The brightness and clarity of that afternoon was stunningly beautiful. Devil's Throne is made of monzonite porphyry, a kind of magma rock that had cooled slowly deep inside the earth as it pushed upward, and also horizontally between layers of Late Cretaceous Mancos Shale, creating what are known as "Christmas Tree Laccoliths," while creating the Cerrillos Volcano. Devil's Throne is the closest Cerrillos Hills formation to the Galisteo River. The Galisteo washed up against Devil's Throne prior to the building of the railroad, which, though not discernible in this photograph, runs along Devil's Throne's cliff face. The two hills to the right in this photograph are very close to the village of Cerrillos. Grand Central Mountain can be seen in shadow in the background.

Galisteo Bosque at Sunset, Autumn. This photograph was taken from Devil's Throne. In the previous photograph of Devil's Throne, I would have been standing on the curve just to the right of Devil's Throne's highest point, located at the left end of the top of formation.

PREHISTORIC PUEBLOS IN AND NEAR THE GALISTEO BASIN

Pueblo San Cristobal, the ancient and elegant habitation farthest to the eastern, upper portion of the Galisteo Basin, was constructed with a greater use of stone than most of the others. At four or five stories high, San Cristobal was, as I mentioned earlier, one of the largest pueblos in the southwestern United States. For Nels C. Nelson it was his first and most extensively excavated site. He excavated 227 rooms.

The abundant, diverse petroglyph art at San Cristobal is regarded as world class. The styles of the petroglyphs range from Puebloan back to Archaic periods.

Built along San Cristobal Creek, there is evidence of water features, such as a dam on one of the many streams in the area, and cultural sites where people may have sat together.

Another feature at San Cristobal is a long wall made of large sandstone slabs that extends completely across the mesa on the east side of the pueblo. This wall may have been built to indicate to visiting groups coming from the east the limit of their approach, and where they may camp. A series of Tipi rings were identified outside this wall by R. W. Lange during his 1976–77 survey.

Pueblo Galisteo, on the other hand, was the last and least excavated by Nelson, but is considered the signature site of the Galisteo Basin. Although there were some stone base alignments at a few of its walls, it was constructed almost entirely of adobe. These adobe mounds are quite tall: 2.5 meters, or eight feet high, indicating that these buildings ascended several stories.

On the Galisteo Volcanic Dyke above the pueblo is Pueblo Galisteo's ancestral site, Pueblo Las Madres, which was constructed of beautiful sandstone. It had 60 rooms.

Pueblo Galisteo was built near to where the long, volcanic dyke that traverses that area dips into a low arc, allowing upstream soils in Galisteo Creek to build up, then spill over, creating an alluvium of very fertile soil. Pueblo Galisteo consisted of 26 room blocks.

Pueblo Colorado was so named by Nels C. Nelson because of the beautiful, red sandstone cliffs and building blocks nearby. It was built near the base of the 250-foot-tall escarpment within the Eastern Mesa highlands, near the Arroyo de la Jara, and included a C shaped water feature that was likely an agricultural reservoir. And it was near a side canyon with springs, near the base of the sandstone cliffs.

Tanoan people living today at Kewa Pueblo call the site of Pueblo Colorado with the name Tze-man Tuo, "place where the eagle's claw is inside."

Pueblo Colorado includes a component of early Archaic occupation consisting of chipped obsidian, basalt, and rhyolite, with a corner notched dart point. Just to the southwest of this Archaic site is a late Developmental habitation (1075 AD to 1175 AD) containing six to ten jacal rooms along with one or two pithouse structures.

Pueblo Colorado during the Classic period, from about 1325 to 1500, consisted of eleven room blocks with 887 rooms, several plazas, and some kiva depressions. The site was occupied during Archaic, Developmental, and Classic periods.

Pueblo She is located downstream from Pueblo Colorado along the Arroyo de la Jara. Nelson documented fourteen room blocks and estimate that there were 1,543 ground floor rooms, nine enclosed plazas and at least four kivas. These structures date from 1325 to 1500 AD, but other, associated locations indicate earlier habitations, at least from 1275 or earlier.

As with other Galisteo Basin pueblos sites, shrines, reservoirs, and agricultural areas were also identified. And, like San Cristobal Pueblo, there is a long, stone wall along the east side of Pueblo She, built of large sandstone blocks. It is half a kilometer long. This wall appears designed to limit the access of people traveling in from the plains.

Trade between the pueblos and the tribes who hunted buffalo out on the eastern plains and beyond was apparently very lucrative for both cultures, but dangerous for the pueblo people.

Just a few miles to the northeast of the Galisteo Basin is the Pecos River Basin, within which Pueblo Cicuye, (now commonly known as Pecos Pueblo and located today in the Pecos National Monument), was built like a fortress along Glorieta Creek. This was because the people of Cicuye specialized in this trade with the Native People of the plains.

There were two rectangular, rock walled structures, each about five stories high. On the inside the rooms had covered balconies up to two levels, which looked out onto a large, central plaza, as well as kivas and other plazas within. There were also tunnels leading under these structures and into smaller structures outside. All of this was surrounded by a low stone wall.

The site of Cicuye, at the southern end of the Sangre de Cristo Mountains, which is called Glorieta Pass, looks out onto the vast, eastern plains. It was the ideal location from which to conduct this trade.

The Tanoan people of Cicuye were part of the Tanoan pueblo culture. However, because their lifestyle was invested in trade with the nomadic tribes of the plains (including the Kiowa people, who were also part of the Kiowa Tanoan language family), the people of Cicuye became very practiced in the arts and customs of both cultures.

With hundreds of warriors always at the ready, and large stores of corn, beans, squash, hides, and feather robes, which were made by twisting together feathers and strands of cotton yarn from which they made an excellent cloth (this technique was also used to make their *mantas*, with which the people protected themselves from the cold), the people of Cicuye commanded the trade between pueblo farmers and the nomadic hunting tribes of the buffalo plains.

Because of its location at an altitude of 6,926 feet, the people of Cicuye were not able to grow cotton or raise turkeys. It is likely that the Galisteo Basin pueblos, along with pueblos all along the Rio Grande, provided most of the local goods for trade at Cicuye.

Pueblo Largo had eight rectangular great houses of stone and adobe, four rectangular kivas, five plazas, and a shrine. It is estimated to have had 480 rooms. Nelson referred to it as the smallest of the Galisteo Basin pueblos. It was built on a promontory with dramatic views of the eastern plains. It dates from the mid-1200s to the mid-1400s.

Pueblo Blanco abuts the north side of the magnificent Crestone Volcanic Dyke, or El Creston, located near the southern boundary of the Galisteo Basin.

Nels C. Nelson thoroughly documented Pueblo Blanco in 1912. He identified fifteen or sixteen separate buildings that were often joined together at the corners, creating six completely enclosed plazas, and six more that were partially open. He estimated 1,450 ground floor rooms, half of which he thought supported upper stories. Nelson's estimated total room count for Pueblo Blanco was more than 2,000.

At Pueblo Blanco people built three large agricultural reservoirs, two of which lay on the opposite side of the formidable Creston Volcanic Dyke from the Pueblo.

Pueblo San Lazaro was declared a National Historic Landmark in 1964. It consisted of 17 room blocks, five midden areas, and a reservoir. It is described as "an exceptional site with extraordinary informational potential." It was first occupied in the early 1300s until the 1500s, and then was reoccupied from approximately the 1600s until the Pueblo Revolt in 1680. Nels C. Nelson estimated San Lazaro Pueblo to have had 1,941 ground floor rooms, and other archeologists have estimated that the pueblo had its total number of rooms at perhaps 5,000.

San Lazaro artifacts from Nelson's excavations are now in the American Museum of Natural History in New York City.

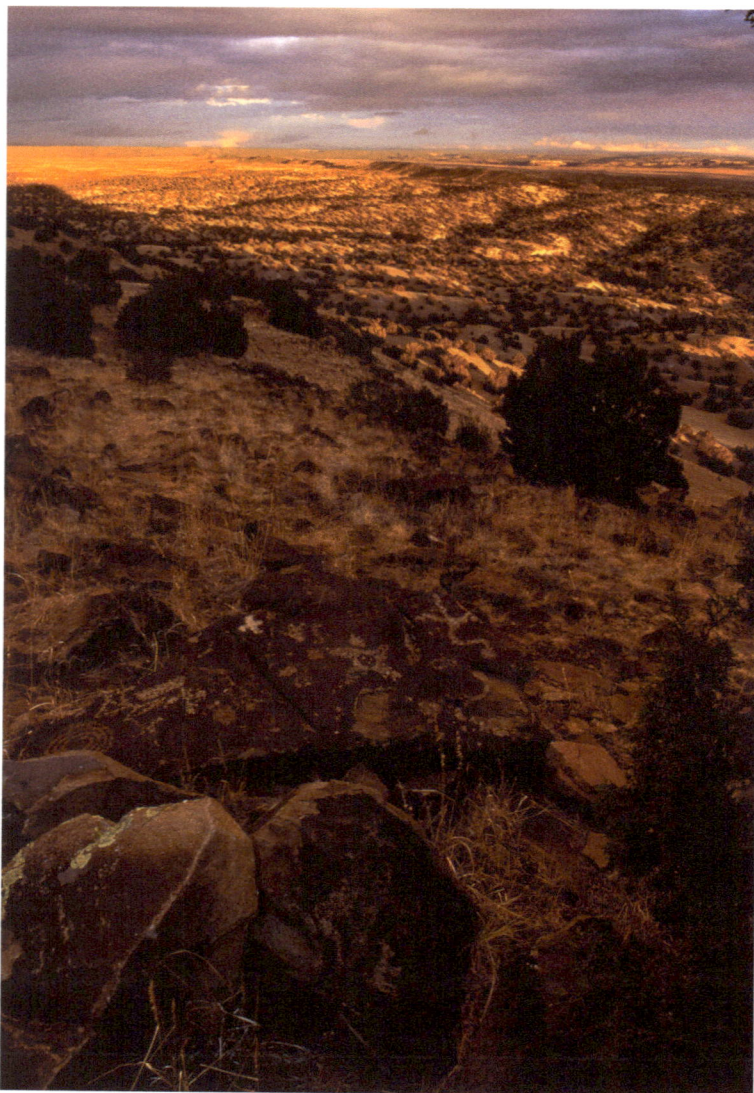

Petroglyph Hill at sunset. Here, looking east from the summit of Petroglyph Hill, part of the Galisteo Volcanic Dyke can be seen as a black line running somewhat diagonally across the background in the upper part of this photograph. Galisteo Pueblo was built near this very long dyke. The Native American petroglyphs in the foreground are very close to the petroglyphs on this book's cover photograph.

There were other major pueblos in and near the Galisteo Basin that were directly a part of this Galisteo Basin Anasazi Civilization phenomenon that are not mentioned here.

The occupation of Pueblo San Marcos dates from the mid to late 1200s AD. Nels C. Nelson conducted major excavations of the site, which he and others estimated to contain 22 room blocks, ten or more plazas, and ten kivas, occupying 60 acres, with extensive agricultural sites nearby. These included rock bordered fields, mulch fields, and check dams.

Rio Grande Glazeware Pottery has been given a date of 1310 AD for its appearance in Anasazi Culture. It is made by rubbing lead ore on the pottery before firing, creating a shiny, black glaze. At first it was used for large pots, which families would bring, filled with food, to feast day celebrations. It is believed that the new pottery style may have been associated with an increased emphasis upon, or changes to, social and religious observations. The Black on White pottery continued to be used for daily purposes, but was gradually replaced completely by the Rio Grande Glazeware.

Anasazi Black on White pottery had been made for a long time. Its white surface was beautifully decorated with precisely drawn, artistic black glaze designs, the glaze made from plant resins. As I mentioned, the Anasazi had developed a system for firing the pottery in a low oxygen environment which was necessary to protect the black glaze from fading as a result of oxidation. Rio Grande Glazeware, on the other hand, with its mineral based glaze, was fired in a high oxygen environment.

The new Glazeware pottery also contrasted with white ware because of the use of a wider range of colored slips. There were a lot of bright red, as well as orange, pink, yellow, and off-white colored pottery

An abundance of turquoise and lead glaze pottery was found at Pueblo San Marcos. Analysis of glazeware found at other Galisteo Basin pueblos indicate that it originated at San Marcos. Pueblo San Marcos is located in the San Marcos Arroyo, very close to the Cerrillos Hills where there are significant deposits of both the sacred turquoise and galena. Galena was the mineral used to create the lead glaze for Rio Grande Glazeware pottery.

It is believed that Pueblo San Marcos was a major manufacturing and trading center for these items, and, it is believed, controlled this trade throughout the Galisteo Basin, and into Native American trading networks in every direction.

Two different language groups built pueblos in the Galisteo Basin. They were the Tano speaking people and the Keresan speaking people.

The Tanoan people may have been living in the Galisteo Basin for a very long time prior to the construction of the pueblos. Some scholars believe that their ancestors have inhabited the Rio Grande Valley for at least two thousand years. The drainage at San Cristobal Pueblo has evidence of habitation from between AD 400–600. Some Basin pueblos exhibit early first construction from the 1000 and 1100s. Including San Cristobal, the Tano built at least seven large pueblos in the eastern part of the Galisteo Basin.

The Keresan speaking people came from Acoma, from the southwest. They arrived in the Galisteo Basin during the 1200s AD. They built Kewa Pueblo in the western portion of the Basin. They probably built Pueblo Blanco and may have re-inhabited the formerly Tanoan site of San Lazaro Pueblo. They certainly built San Marcos Pueblo. San Marcos Pueblo was about two miles from the turquoise mines in and near Mount Chalchiuitl and five miles from the turquoise mines in the hill called Chalchiquite.

Though culturally similar, the languages of these two groups are not related. The Tano language is part of the Kiowa-Tanoan language family. Although it is not known for certain, at least some of the people at Mesa Verde may also have spoken Tanoan, possibly related to the original Tanoan people of the Galisteo Basin.

The Keresan language is a language isolate that originates from Acoma. Acoma, along with the Hopi village of Oraibi, both probably settled in the 1000s, or much earlier, are the two oldest continuously inhabited communities within the United States. Chaco Canyon lies between them and to the north. The Hopi speak a Shoshonean language of the Uto-Aztecan language family. The Hopi village of Orayvi, on Third Mesa, is thought to be the oldest continuously occupied settlement in the United States.

The Keresian speaking people may also have been related through language and/or through trade to the people that lived at the Rio Puerco/Guadalupe Community.

Guadalupe, the eastern most Chacoan Outlier, was built near the Rio Puerco drainage upon an isolated 200-foot-tall sandstone mesa that had sheer walls on all sides, with dramatic views of deep canyons, rugged mesas, and volcanic formations. It was located due west of the Galisteo Basin and approximately midway between the Galisteo Basin and Chaco Canyon, on the northeast side of the 11,305-foot volcanic peak, Mount Taylor. Even today the Guadalupe Ruins exist in a very remote location.

No one knows for sure what language or languages were spoken at Chaco Canyon. The people of the Guadalupe community may have been responsible for transporting the turquoise from the Cerrillos Hills to Chaco Canyon. Guadalupe was inhabited early and was used for a long time. Much turquoise was found at Guadalupe.

Mount Taylor is also close to, and particularly sacred to modern day Acoma.

Most of the Turquoise found at Pueblo Bonito in Chaco Canyon was sourced to have come from the Cerrillos hills, creating speculation that this Great House had its own turquoise importation network. The ancestors of Keresan Cochiti Pueblo, just north of Kewa, built Tyuonyi Pueblo in Frijoles Canyon around 1350, as well as other beautifully constructed nearby sites. They were constructed from, and within, the tuff boulders and cliffs in what is now Bandelier National Monument along the Pajarito Plateau in the Jemez Mountains.

The Pajarito Plateau is made of a layer of tuff, which is compressed volcanic ash, and a layer of basalt, or lava, that lies beneath this tuff. The volcanic ash was deposited by Jemez eruptions from 1.61 and 1.22 million years ago. Tuff is a relatively soft material, ideal for construction of the cliff dwellings that were built along the walls inside Frijoles Canyon.

Conflicts with their Tanoan neighbors, ancestors of present-day San Ildefonso Pueblo, and possibly other nearby Tanoan pueblos, forced the Bandelier people to eventually abandon these Jemez Mountain homes and move down to the Rio Grande to present day Cochiti Pueblo, north of Kewa Pueblo, and to San Felipe Pueblo, south of Kewa.

Here is a quote from a letter written by my friend; the artist Douglas Johnson, to his friend, Mike Bremer, head archeologist for The Santa Fe National Forest:

"If Chaco's great houses were ceremonial structures, then the dances at Kewa originated there. I thought of the Chaco road system as I waited an hour and a half to get into the village to park. So many had come from afar to attend. An annual pilgrimage. And the spectacle of dance and ritual and the feeding of thousands must have roots in Chaco. And I trace the migrations of Queres clans from Chaco to the Parijito Plateau, and to the Rio Grande Valley to Kewa. The trade in turquoise and Heishi, the beads of shell and turquoise, came from Chaco. And only Kewa has continued the tradition. I can remember people still bow drilling beads and grinding strings of beads on stone when I first arrived there in 1970. So Kewa has become the new Pueblo Bonito. As for secrecy and tradition Kewa is the most intact, more than Hopi. We are lucky to live and see such so close."

I have a hypothesis. The Tano speaking people had lived in the Galisteo Basin for a long time, at least from the Archaic Period. They had practiced their agriculture, their ceremonies, and other Anasazi customs within the Basin for hundreds, possibly thousands of years. The Keres speaking people, although of a different language group, were also culturally Anasazi, but moved to the Galisteo Basin with an additional motive: They established San Marcos Pueblo specifically to secure or regain control of the turquoise mines nearby, and to continue or re-establish control of the turquoise trade after trading networks centered at Chaco Canyon had disappeared.

South, in what is now Northern Mexico, Casas Grandes, also known as Paquime, was founded by people of the Mogollon Culture at about 1130. The people there developed sophisticated irrigation systems that resulted in abundant crop yields. By 1350 the agricultural center, Casas Grandes, had become a major trading center in Northern Mexico, between the Aztec Empire to the south and the Mogollon and Anasazi Civilizations to the north. They specialized in raising macaws and in the manufacture of copper bells.

Casas Grandes likely became an important trading partner for San Marcos, as well as other Rio Grande area pueblos. Although many unrelated language groups of many seemingly isolated cultures of prehistoric Native Americans existed across the vast landscape of North America, there had also been, for many thousands of years, trading networks in every direction that connected these groups. And this trade was a common, consistent, and important feature for all of them.

One of the early pueblos in the Galisteo Basin was Burnt Corn. Burnt Corn Pueblo was built on a ridge top above a spring near Petroglyph Hill, about three miles southeast of San Marcos Pueblo. Tree ring data indicates that Burnt Corn was built between 1290 and 1302. At about 1310 it was intentionally and thoroughly burned from both the inside and outside of the structures. No one knows for certain who did this or why it was done. But the occupations of Burnt Corn and San Marcos almost certainly overlap, and there would have been some kind of relationship between the two.

Several miles directly behind Burnt Corn Pueblo, to the east and southeast, stood the ancient Tanoan Pueblos, such as Galisteo, San Cristobal, She, Colorado, and Largo. And to the northeast, up Galisteo Creek and near to the Sangre de Cristo Mountains, Manzanares Pueblo was built during a period around the year AD 1295, the same time frame as for the construction of Burnt Corn. Pottery samples collected at Manzanares compare attributes shared by Galisteo Black on white and Mesa Verde Black on white ceramics, which led archeologists to conclude that Manzanares Pueblo was built by migrants from the Mesa Verde Civilization.

During the time when Burnt Corn Pueblo was being constructed there was a convergence of pueblo groups in the Galisteo Basin. Keresan immigrants came from the west during the early to mid-1200s, and immigrants from Mesa Verde came from the northwest during the late 1200s. Both of these groups converged within the Galisteo Basin near to its ancestral Tanoan residents.

Burnt Corn Pueblo was built in a central location between these pueblo groups, and near a water source that allowed it to be as close as possible to the sacred site of Petroglyph Hill, much closer to it than any other Pueblo had ever been.

As droughts caused the abandonment of Chaco Canyon in northeastern New Mexico, and then Mesa Verde in southwestern Colorado, the Galisteo Basin, and the Rio Grande Valley in general, received consistent rainfall.

Other evidence of burning during the same period at other locations within the Galisteo Basin also exist: Cholla House and Slope House are in the vicinity of and associated with Burnt Corn Pueblo. But the three small Lodestar Community sites are located farther away on a sandstone mesa above the south bank of the Rio Galisteo, south of the Cerrillos Hills, and just south of the modern-day village of Cerrillos. Found within the burnt remains of structures at each of these areas were pieces of turquoise, turquoise beads, tools that may have been used to work the turquoise, and obsidian projectile points.

Although there are other hypothetical interpretations for this evidence, recent excavations and investigative research at Burnt Corn and other nearby Galisteo Basin pueblo sites by the Tano Origins Project indicate violent conflict. Burnt Corn Pueblo was completely destroyed by fire while recently harvested corn was drying on its roofs, probably between 1302 and 1310 AD. It was never re-inhabited. (See *Burnt Corn Pueblo, Conflict and Conflagration in the Galisteo Basin, A.D. 1250–1325*, James Snead & Mark W. Allen, Editors, University of Arizona Press, 2011.)

Eagle Dancer Petroglyph

The Eagle Dance is performed today by Native American tribes all across North America, by tribes such as Iroquois, Delaware, Hopi, Zuni, Sioux, Cheyenne, Comanche, and Pawnee, just to name a few. Near the Galisteo Basin the dance is particularly significant to Jemez and Tesuque Pueblos. Tribal groups believe that the Eagle serves as a messenger between humans and the Creator, and that the Eagle has the power to control rain and thunder.

About 1400 AD a new group of people appeared in the vicinity of the Galisteo Basin: the Dine'. This Athabascan speaking people had apparently migrated into New Mexico and Arizona from northwestern North America. They are the ancestors of modern-day Apaches and Navajos. They traded with and raided the Galisteo Basin Pueblos. This influx probably entered the Basin from the east, from the plains, and so the pueblos of the upper Basin may have been most impacted at first. Several of these pueblos were abandoned by the early 1500s and the people of San Marcos Pueblo built their great defensive compound at this time. All the pueblos of the Galisteo Basin were severely stressed by the Dine' over the course of that century. It has been postulated that there are other tribal groups, beside the Dine', who may have been involved in raiding the Galisteo Basin Pueblos during this period.

Francisco Vasquez de Coronado passed through the Galisteo Basin in 1541, making first European contact, and beginning the Basin's written history, as he began his journey north in search of Quivira, the fabled "Cities of Gold." His chronicler, Pedro Castaneda, described three Basin pueblos which had been severely depopulated by an attack sixteen years earlier. Coronado visited Galisteo, San Lazaro, and possibly San Cristobal Pueblos. Coronado did not visit San Marcos Pueblo.

Coronado's men carried crossbows. Several crossbow bolt points made of copper have been discovered in New Mexico. One bolt point was reportedly found in the Galisteo Basin. Because the find was taken without proper archeological documentation it cannot be verified. But the point is said to have been found at San Lazaro Pueblo on the Arroyo Chorro. By the time the next Spanish incursion entered the Galisteo Basin forty years later crossbows were obsolete, having been completely replaced by firearms. The Rodriguez-Sanchez Chamuscado entrada was assembled to spread Christianity, but some members were interested in something else.

When at San Marcos Pueblo, "we asked if there were many minerals in the vicinity. They consequently brought us a large quantity of different kinds, including some of a coppery steel like ore," according to *The Gallegos Relation of the Rodriguez Expedition to New Mexico 1581-1582*, translated and edited by George P. Hammond and Agapito Rey.

Besides turquoise the San Marcos people mined lead for their lead glaze-decorated ceramics. This lead ore, or galena, was traded throughout the Southwest, as was their pottery, now famously known as Rio Grande Glazeware. Evidence suggests that other tribes traveled long distances to the Cerrillos Hills, passing up other known sources, specifically to obtain the Cerrillos Hills galena, possibly because the galena mines, located very near to the turquoise mines, were considered sacred.

Two of the Spaniards in the expedition were familiar with galena ore and knew that it contained silver.

It was when people from San Marcos Pueblo brought them to their galena mines in the Cerrillos Hills in 1581 that these Spaniards mined the first silver in the United States. Of the three priests with the entrada, one decided to leave the Basin by himself, bound for Mexico. The expedition then traveled east to look for bison. Upon their return to San Marcos, they were told that Frey Santa Maria had been followed by some Tano men who killed him a few days later while he slept, dropping a large rock on him, a death reserved for evil witches.

The other two priests chose to remain behind when the entrada returned home, at Puaray Pueblo near modern day Bernalillo. That pueblo had been devastated forty years earlier by Coronado and his men. The priests were soon killed.

In 1590 the Castano de Sosa Expedition traveled north from Mexico looking for the source of minerals taken to Mexico by the Rodriguez-Sanchez Chamuscado entrada, where they had been assayed and found to contain silver. In January 1591 the de Sosa Expedition established a settlement at San Marcos Springs in the San Marcos Arroyo, just south of San Marcos Pueblo. In doing so they established the northern third of El Camino Real from there.

Early Spanish explorers traveling north had favored turning east, up the Galisteo Basin, from their route up the Rio Grande River. This was because of the formidable barriers to travel created by the imposing basalt cliffs of La Bajada Mesa and Escarpment, and the gorge of the Rio Grande, where the river passes through the eastern flank of the Jemez Mountains.

The DeSosa Expedition abandoned their site at San Marcos Springs after a few weeks, however, and traveled back down the San Marcos Arroyo, past the site of the modern-day village of Los Cerrillos, located at the juncture of the San Marcos Arroyo with the Galisteo River.

One of the two men who were taken to the Galina mines in the Cerrillos Hills by men from San Marcos Pueblo during the Rodriguez-Sanchez Chamuscado entrada in 1581 was Felipe de Escalante. He returned to northern New Mexico with Juan de Oñate in 1598, and Escalante showed these mines to Oñate, who later referenced them as, "las minas de Escalante."

Juan de Oñate, who was married to a granddaughter of Herman Cortez and an Aztec princess named Isabel Moctezuma, is sometimes referred to as the Last Conquistador. He established New Mexico's first Spanish Colony with 400 settlers in July 1598, an event that, in time, would greatly impact the Galisteo Basin.

This first colony was located at the juncture of the Chama River with the Rio Grande, just north of modern-day Espanola. The Tanoan speaking people of Ohkay Owingeh and Yunque pueblos were removed from their homes. The Spanish Colonists then remodeled these Native pueblos into dwellings that were more familiar to themselves, such as by placing windows in the ground floor rooms.

The colony did not prosper. The people were unable to grow enough crops, and the expected silver lodes did not appear. Many settlers wanted to return to Mexico, but Oñate refused to allow them to leave and executed some of their leaders.

Juan de Oñate's father had been renowned for establishing silver mines in Mexico, and so, as one might expect, this son's primary motive for colonizing New Mexico had been to discover gold and silver.

Felipe de Escalante was killed in early December, 1598, along with Oñate's nephew, Captain Juan de Zaldívar, and ten other Spanish men at Acoma Pueblo.

The Acoma chief, Zutacapan, had sought to negotiate with the Spanish to avoid violent conquest. Oñate had sent his nephew to Acoma.

Upon their arrival Zaldívar took sixteen of his men up onto the Acoma mesa stronghold. Initially the discussion between Zaldívar and the Acoma people was peaceful. Then Zaldívar began demanding grain from Acoma's winter stores, which the people needed to endure the winter months. There then appears to have been a conflict between some Spanish soldiers and some Acoma women. Acoma warriors attacked. This event became known as the Acoma Revolt.

In late January 1599 Oñate sent Juan de Zaldívar's brother, Vincente de Zaldívar, to Acoma. For two days the two sides, native people defending their home and the Spanish Conquistadors who attacked them, skirmished inconclusively.

Acoma is built upon a 357-foot-high sandstone mesa. The Acoma people were known as "the people of the white rock." For two days Acoma Warriors held out the Spanish invaders, throwing rocks and shooting arrows down the cliffs at them. But Zaldívar, with twelve men and a small cannon, managed to climb the back of the mesa. On the third day they began firing the cannon into the fortified pueblo. After a period of time the walls of the pueblo were breached and several Acoma homes caught fire. The Spanish invaded the village. Eight hundred Acoma people were killed, three hundred of them women and children. This event became known as the Acoma Massacre.

Survivors were rounded up and marched, in the middle of winter, to Kewa Pueblo, whose people are relatives of the Acoma people, in order to stand trial. Boys over twelve years of age were sentenced to twenty years of servitude. Men 25 years and older were also sentenced to twenty years of servitude and to have a foot cut off. Finally, 24 of these men did have a foot cut off, and two Hopi men, who just happened to be visiting Acoma at the time, each had a hand cut off. Females above twelve years of age were also sentenced to twenty years of servitude. Sixty young girls were deemed not guilty and sent to Mexico City to live in convents. But some Acoma people escaped and went home. By 1601 Acoma Pueblo was completely rebuilt.

In 1600 Vincente de Zaldívar claimed that he had discovered the Cerrillos Hills mines, the very mines that had originally been revealed to Felipe de Escalante and another man in 1581, by friendly San Marcos men who were almost certainly Keresian speaking relatives of the Acoma people.

Juan de Oñate made significant discoveries during expeditions that he undertook. In 1601 he traveled eastward, onto the Great Plains with 130 soldiers and 12 priests. Like Coronado before him, he was looking for Quivira. He followed the Canadian River into present day Oklahoma where he met with the Dine' people.

During this expedition Oñate made the very first chronicle describing the Tall Grass Prairies of North America. He chronicled his encounters with several other tribal groups of buffalo hunting people including the Escanjaques who Oñate noticed wore buffalo skins. Oñate also took hostages, some women and children, and a tribal chief of the Rayados tribe named Catarax whom he utilized as a guide. Oñate greatly respected the Rayados people, and especially Catarax whom he treated very well. He released the hostages on his way home, though he changed his mind about some of the children at the request of his priests, who wanted to convert them to Christianity.

When he returned to his colony, already dejected for having found no treasure, Oñate discovered that most of the colonists had left during his absence. A few colonists, who had been loyal to him, remained.

During 1604 and 1605 Oñate, with about three dozen men, traveled from the Rio Grande all the way to the Gulf of California, exploring much of the Colorado River. He visited Zia Pueblo and the Hopi, and many tribes to the west, including the Mojave and Pima.

When King Philip III of Spain eventually heard about the Acoma Massacre, he summoned Juan de Oñate to Mexico City.

Oñate finished his plans to re-establish the New Mexico Colony at Santa Fe, which displaced more native people, and then, unaware that the king intended to file charges against him, he resigned as Governor of New Mexico in 1607, and traveled to Mexico City. Juan de Oñate was tried and convicted of cruelty and banished from New Mexico. He returned to Spain, where he spent many years trying to reverse this conviction. He eventually succeeded. He was later appointed head of all Spanish mining inspectors by the King of Spain.

The Native inhabitants that were then living in today's American Southwest grew increasingly angry over their treatment by the Spanish after the Spanish Colony was re-established at Santa Fe.

There were layers of complexity to this problem: excessive tithing and tributes of their food and clothing, forced labor, enslavement, and executions. Then a prolonged drought, beginning in the 1660s and worsening in the 1670s, had many consequences. These included an increase of attacks by the Dine' (Apaches) on Spanish and pueblo settlements, so that the Spanish lost much of their capacity to protect these pueblos. But foremost was the Spanish preventing the pueblo people from practicing their traditional ceremonies, and the destruction of their kivas. Although the pueblo people had accepted Christianity into their communities, the Franciscan priests wanted to completely eliminate their ancient traditions.

In 1680 a majority of tribes that included the Tano, Keres, other pueblo language groups, and Dine', joined forces and drove the Spanish out of New Mexico during what became known as the Pueblo Revolt. Many priests at pueblo missions were murdered. The pueblo people of the Galisteo Basin and the Apaches joined in the attack on Santa Fe.

There was a large ranch headquartered in the San Marcos Arroyo which passes through the eastern edge of the Cerrillos Hills. It was owned by a man named Sargento Mayor Barnabe Marquez, and was most likely located at the San Marcos Springs, less than two miles down the San Marcos Arroyo from San Marcos Pueblo, probably at or near the sight of the Castano de Sosa expedition settlement.

Many of the Spanish colonists from nearby gathered there. They endured a large and sustained attack on this ranch. Everyone was able to escape with their lives, however, during the night. Sargento Mayor Barnabe Marquez remained in Mexico with his family after the Pueblo Revolt, not returning to New Mexico.

During the revolt the Native people buried and hid the Mina del Tierra and Bethsheba mines in the Cerrillos Hills. These had been their ancient locations for procuring galena for their Rio Grande Glazeware Pottery. The Spanish had taken over these mines and then used the Native miners as slave labor. It has been reported that the Revolt of 1680 came about after several Native slaves working the mines were buried alive in a rock slide.

One very significant consequence of the Pueblo Revolt resulted from the thousands of horses left behind by the Spaniards' hasty disappearance. Most of these horses were released by the pueblo people, who had little use for them. This event became known as the Great Horse Dispersal.

The Iberian Mustang, brought to the new world by the Spanish, was very different from other, larger, European horse breeds. It had been brought to Spain by the Moors from North Africa. Ancestors of these North African horses had migrated there from the steppes of central Asia, by way of Arabia and the Middle East. And the ancestors of the steppe horses had, thousands of years earlier, crossed into Asia through Beringa, from North America.

Mare and Colt. I saw this mare and her colt when I was photographing on top of La Bajada Mesa. They didn't want me to get too close, so I used my 400 mm lens. The light happened to be perfect for my slide film.

These Spanish Mustangs thrived, building wild herds, re-establishing the equine, which had gone extinct perhaps 10,000 years earlier, back into the prairies of the American west.

Native American tribes quickly took advantage of this new resource. One of the first and most successful was the Comanche. The Comanche people were adept at capturing, training, and breeding horses. Perhaps their capability as horsemen was related to their past. They had been a small group of Shoshone speaking, very primitive, hunter gatherers who had been marginalized into portions of what is now eastern Wyoming by their more powerful neighbors. These marginalized people, after becoming mounted, expanded their territory into what became the largest Native empire in the history of North America, encompassing parts of Wyoming, Nebraska, Colorado, Kansas, Oklahoma, New Mexico, Texas, and into some of northern Mexico. During the 1700s the Comanches began raiding pueblos and Spanish settlements in New Mexico.

When the Spanish returned to Northern New Mexico twelve years after the Revolt they found all of the Galisteo Basin Pueblos, other than Kewa, abandoned. Many of the people had moved to Santa Fe. Some of the Tano Speaking people had moved to Kewa, while others had traveled all the way to Hopi, where their descendants live still, and continue to speak their native Tanoan language.

Meanwhile, pueblo people had fortified themselves on at least three mesa top villages: at San Ildefonso, and near Cochiti and Jemez Pueblos.

After a bitter, bloody battle the Spanish, under Don Diego de Vargas, were able re-take Santa Fe in 1693, and then re-establish their Northern New Mexico Colony.

In 1695 Governor de Vargas appointed Alfonso Real de Aguilar as alcalde of the mining camp at El Real de Los Cerrillos, a location on the north side of the Cerrillos Hills which is now called Bonanza Creek. Because of this, the Cerrillos Hills became the first documented mining district in the American west. The purpose of El Real de Los Cerrillos was most likely to re-open the Mino del Tierra and Bethsheba Mines.

El Real de Los Cerrillos was abandoned and probably destroyed during the Pueblo Revolt of 1696. This revolt was launched by fourteen pueblos, with the executions of five missionaries and 34 settlers, their response to Spanish efforts to eradicate their Native culture and religion. Though it was short-lived, this Revolt did win the pueblos some major concessions from the Spanish: Substantial land grants were issued to each pueblo, along with a public defender to protect Native American rights in Spanish courts.

In 1706 Governor Cuevo y Valdes helped the people from Galisteo Pueblo, who had been living at Tesuque Pueblo, re-establish themselves at their Galisteo Pueblo home. But, in 1794, Comanche raids and a smallpox epidemic forced the people to leave. They also took up residence at Kewa.

This, except for Kewa, was the end of over five hundred years of Native American settlement in the Galisteo Basin during the Coalition and Classic Pueblo Periods, plus thousands of years during earlier developmental periods.

The few remaining people at Pueblo Cicuye finally abandoned their homes in 1838, after more than 500 years of occupation, returning to live with their Towa-Tanoan speaking relatives at Jemez Pueblo. This decline and final abandonment of Cicuye had also been exacerbated by relentless Comanche raids and smallpox epidemics.

Kewa and the other the pueblos north and south along the Rio Grande and its tributaries, retain, to this day, their distinctive Anasazi Culture, which can trace its roots back thousands of years, through the Archaic Culture, and before that, to the Paleo Indians, and back through time to the earliest cultures of mankind upon the earth.

The people dance on Feast Days, beautifully costumed in traditional clothing and ornamentation, while ancient rituals are occurring, secretly, inside the Kivas.

And to this day the people of Kewa claim traditional ownership of the Cerrillos Hills turquoise and are renowned for their fine turquoise jewelry.

The artist Douglas Johnson's letter to Annie and Mike Bremer:

"Annie and Mike,

Missed you at Jemez and Kewa. But then you might have been there. There were so many people and I tend to bury myself in a cool shady place among the crowds and stay there all day. When one stays in one place so much more can be seen. And it's all in the details. Jemez waited again until the last dances to bring forth its teen genius of the Turquoise moiety. He drummed to his own compositions with marvelous form and power. He played tones over the surface of the drum, and at one point a new drum was brought in and without skipping a beat he switched drums and the tone went higher. A dancer resting beside me told me the words describe Paradise. I tried to imagine a Pueblo paradise.

"Kewa was huge and introduced some new elements I had never seen before. Four female clowns danced among the male clowns that over the years have come to resemble katsinam. One female clown had the audience in hysterics suggesting sexual antics with a male clown. I wish I could have understood what she said. The Pumpkin moity introduced an innovation by stopping the drum and rattling the rhythm. And during the last dance the drummer stopped drumming and all the dancers were suddenly confused only to have the drum start up again. Everyone laughed and finished the dance.

"After the dance I went to my friend's house to eat and watch the dancers come in from '_ _ _ _ _ ' , the ritual bathing after dancing in the Rio Grande. Grandma blessed them as they came in and each presented her with an evergreen. The house was full. And then the rain began. In a wild fury everyone whooped for joy and rushed outside to stand in the rain. I went with them and everyone held their arms to the sky and held up their babies to receive the blessing of the downpour. As men came in from the kiva, they were thanked for a ceremony appreciated by 'those above'...."

A dear friend and anthropologist named Stephani Salkeld introduced me to Leo. We stayed the night at his house with him and his family. Leo was an Elder of some kind, and a jeweler. He was a World War II veteran and I remember he had a personal letter to him, framed on his wall, signed by President John F. Kennedy. He also had a ranch on the pueblo, on the other side of the Rio Grande.

I worked for him there once, with Todd and Eric. We all stayed overnight at the little ranch house. Leo paid me three bales of hay. I was kind of disappointed that I was not paid in cash. Leo invited me to come back and work for him again, and he would pay me in cash. I told him that I would have to ride there on my horse. He said that would be fine and we made arrangements for me to arrive on a certain day. On that morning, in May, Galahad and I started down the Galisteo. After a few miles galloping in the sand along the river (creek) we rode up onto the flats on the south side of the river, past the Ortiz Mesa foothills, and then headed in the direction of Kewa Pueblo. I was always good at getting through the fences without cutting any wires. At one place I led Galahad deep into an arroyo so we could pick our way under a fence where it was weighted down in the arroyo bottom with wire attached to rocks.

Later we rode up onto a little mountain, through a pass between its two summits, and then down onto the flats again. It started to rain. We galloped. In a while the rain stopped and the wind came up, drying out my shirt. Finally, we reached the highway between Albuquerque and Santa Fe, crossed it, then rode slowly beside the road to the Pueblo in warm sunshine. At the front of the Pueblo Leo's son, Leonard, was waiting for me, sitting in his car. He jumped out, waving me off. "You have to turn back," he yelled. "I can't," I called back. "It's afternoon. I don't have enough time to make it back to Cerrillos." "The deal's off. You have to go back." So Galahad and I turned around.

I wondered why Leo had changed his mind, but thought maybe it was just as well. Galahad was a stallion and it would perhaps not have been so wise to ride him into the herd of pueblo horses running loose down in the Rio Grande River Bosque, on our way to Leo's ranch. Anyway, non-native people weren't supposed to be on the pueblo at night and parading through on Galahad might have been too sensational.

Where the road intersected with the highway, in the days before it had become an Interstate, there used to be some trash cans. I got down and looked in them for food. Two pueblo women in a car saw what I was doing, drove up, and gave me a loaf of their horno baked Indian bread. Galahad insisted I share it with him.

Crossing the highway, we went down the dirt road on the other side, looking for the gate to let us through the fence. A man pulled up in his car to admire Galahad. He said to me that this horse was most beautiful horse he'd ever seen. Galahad was well made, muscular with a big, proud neck. But what the man admired most of all was his beautiful, unusual color; a dark chestnut with purple highlights. When it got dark, I staked Galahad out and slept under a juniper tree. It drizzled all night, but I managed to stay dry between my two saddle blankets, using my saddle pad as a pillow. I slept well.

The next morning was cool and cloudy. After riding a while, it began to snow. It was a beautiful May morning with the smell of the snow in the air. We took our time as we picked our way through the canyon and mesa foothills on the west side of the Ortiz.

Back on the river it snowed again. It felt good to be close to home...close to Dorothy. It felt happy and warm. We cantered toward town.

Looking back, I wonder if maybe Leo didn't change his mind. Maybe all he wanted all along was for me to ride Galahad to Kewa. I guess I may never know for sure.

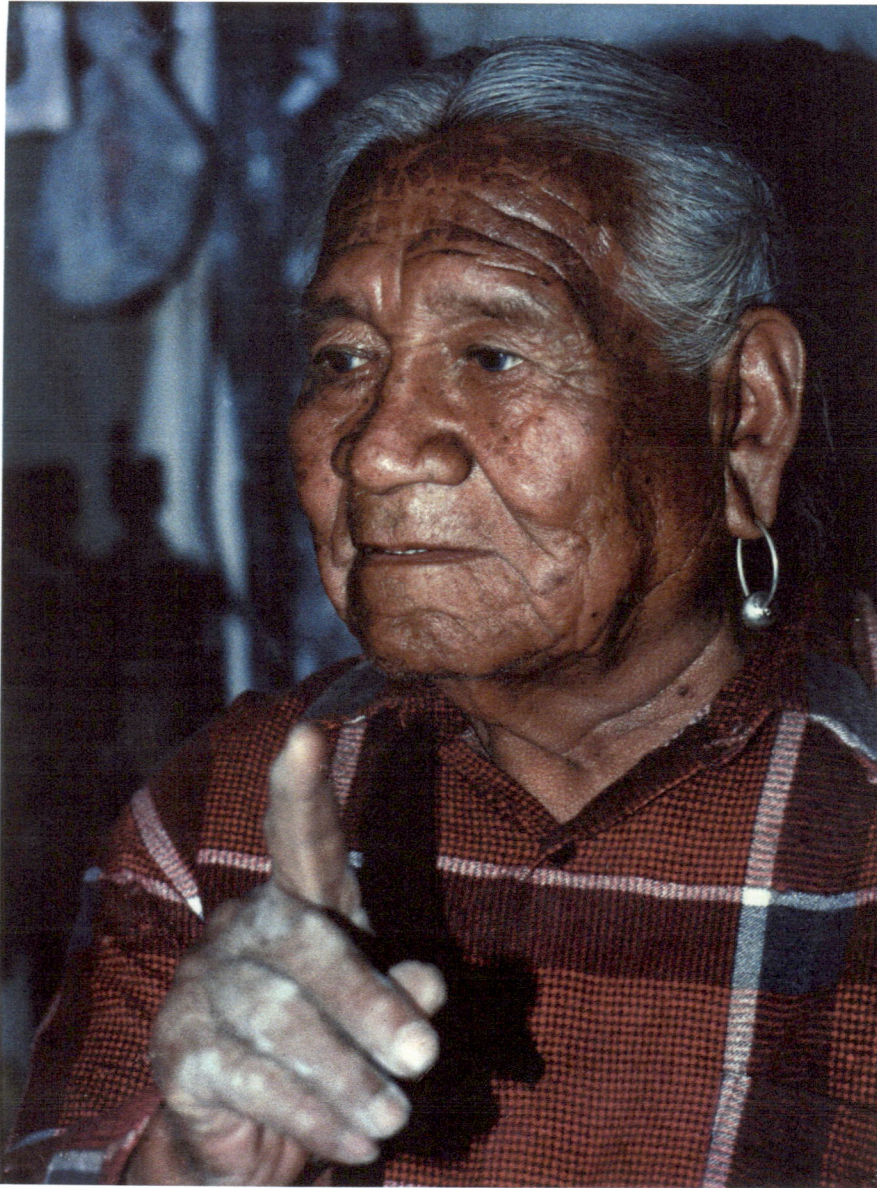

Leo Corez. Leo was a tribal leader, Elder, and traditional jeweler from Kewa Pueblo. Photograph Courtesy of Jeff Salkeld.

The Spanish presence in the Basin was never very large. The village of Galisteo was founded in 1816 with nineteen families.

Some mining did take place in the Cerrillos Hills after the 1696 Revolt, primarily by prominent Santa Fe families. Few records exist. The Galina was mined for both the lead and the silver. Spanish artisans created religious works, such as uniquely designed silver crosses.

From the beginning Spanish colonists were isolated from their origins in Mexico and Spain, and a very different culture emerged. It lives on today among their descendants. Spanish is spoken more slowly, with a different vocabulary and accent than in Mexico.

The phrase, *The Land of Poco Tiempo* (The Land of Slow Time) emerged long ago as the title of a book by Charles F. Lummis (1893) who came to New Mexico in 1888 for his health. His book describes a culture that Lummis discovered here, with "...the lack of haste in the lives of its inhabitants."

Mostly, the Galisteo Basin was used for pasture. In 1810 the San Marcos Grant was denied to Francisco Ortiz because it was reserved for the poor of Santa Fe, so they might have a place to graze their cattle.

Iglesia Nuestra Senora de los Remedios, located in the village of Galisteo. The Village of Galisteo was founded in 1816 with nineteen families. This church was built in 1884. It was constructed upon the exact same dimensional location as the previous church which was built in 1706 by early Spanish colonists, according to The Galisteo Basin and Cerrillos Hills by Paul Secord and Homer Milford.

While the Cerrillos Hills revealed silver, the Ortiz Mountains spilled gold!

Motivation for prospecting and mining by the Spanish was lacking, partly because of the requirement to tithe a significant proportion of any mineral wealth to the king of Spain. Also, trading with the Americans could not occur because Americans were not allowed to enter into the Spanish Colony, and in fact, were imprisoned by the Spanish for doing so.

There was some washing of placer deposits in the Ortiz during the Spanish Colonial period, but it was with the coming of the Mexican Republic and the opening of the Santa Fe Trail, both in 1821, that serious gold mining and the ensuing gold rush in the Ortiz Mountains came to life. No longer would the Americans, with their wagons filled with goods, be kept out! No longer must tithes of silver and gold be paid to the King of Spain!

The gold camp of Dolores would occupy about 580 acres. By the 1830s there were periods during the winter months when up to ten percent of the population of New Mexico was in the Ortiz Mountains looking for gold, according to Bill Baxter, of whom more will be mentioned later.

The gold rush continued well into the American Territorial Period beginning in 1848. The mining boom in the Cerrillos Hills began in 1879, attracting an influx of miners, many from Leadville, Colorado. That and the coming of the railroad in 1880 gave birth to the town of Los Cerrillos, which was such a boom town that the American Territorial legislators seriously considered making Cerrillos the Territorial Capitol, until, some say, major flooding of the Galisteo River and the resultant damage to Los Cerrillos changed their minds. Los Cerrillos was said to have 26 saloons and four hotels.

In 1885 Richard Greene settled in Cerrillos. He and his family had herded their cattle up from Texas with four covered wagons. In 1888 he hired traditional builders from Cochiti Pueblo to help create the Palace Hotel, which became "Cerrillos' grandest and finest hostelry" according to Bill Baxter.

It was constructed upon a home, and possible earlier stagecoach stop, made of adobe rooms, to which were added an elegantly built, forty-four room quarried sandstone and adobe structure three stories high.

The Ortiz Mountains from the Waldo Road. "While the Cerrillos Hills revealed silver, the Ortiz Mountains spilled Gold!"

There were long, tall stone walls, and angled walls with balconies before second floor rooms. Several pitched roofs complemented second and especially third story windows.

The Palace Hotel hosted the town's doctor's office, where Dr. Friend Palmer continued caring for the town's people his whole professional life. The town's dentist office and tailor shop were also located there.

The elegant building became a place where the Cerrillos residents could feel welcome and at home. On the ground floor there was a living room with a table and check in desk; a spacious dining room which had a high ceiling that ascended to the top of the second story over a long and elegant dining table, a buffet stand displaying beautiful china, and a fireplace. In the kitchen there was a large wood burning cook stove, a food preparation area, and another table with chairs for dinners. Everything was lit by kerosene lamps. On the second and third floors each guest room had elegant furniture, a basin and water pitcher made of china for guests to wash their hands and face, and Tiffany lamps on night stands. The outhouse was outside in the back.

The Greene family's eldest son, Clay, responded to Colonel Theodore Roosevelt's call to join the Rough Riders in the western United States. Clay Greene was killed during the charge up San Juan Hill while serving as Colonel Roosevelt's Orderly on July 1, 1898. His grave is in the old Protestant Cemetery in Cerrillos near the graves of other family members, including his father who was buried next to him some years later.

The mining boom also caused the town of Carbonateville to come to life nearby in the Cerrillos Hills. By 1884 it had a population of 500 people. Lew Wallace lived for a time in the Carbonateville Hotel while working on his novel, *Ben Hur*.

St. Joseph Catholic Church in Snowstorm. St. Joseph Church was built in 1922, replacing the original church built in 1884. It was constructed by a local builder named Frank Schmidt who happened to be living at the time in a house directly across River St. from the Church's location. He was affectionately called "Doche" by all the local residents, a local slang term for Dutchman. Frank's grandson, Larry, still lives in the village. The statue of St. Joseph, seen to the left of the front of the church, was carved by an artist from a dead Chinese elm tree. This photograph was taken in late December, 2006, during a snowstorm that produced Cerrillos record snowfall of thirty inches.

During the late 1800s increased cattle grazing reduced the Basin's abundant grass cover. This resulted in the formation of arroyos, and there was major flash flooding of the Galisteo River that devastated Kewa Pueblo. The Franciscan Mission that was there, built in 1607, was one of the largest in New Mexico. The native people tried to save it, but each year the river encroached upon it, until the foundation crumbled and disappeared in 1886.

The people of Kewa relocated their pueblo to the Rio Grande, just below its confluence with the Galisteo River. To this day the people remember and speak of the destruction of their ancestral pueblo.

Then the Cerrillos Hills mining boom played out. Miners began to leave, looking for richer deposits elsewhere. By 1892 Carbonateville had 40 residents and was eventually abandoned. In 1898 Thomas Edison was said to have stayed at the Boarding House Hotel in Cerrillos while visiting the Ortiz Mountains regarding his new idea for a gold separation process. A mill was built in the Ortiz near to the town of Dolores in 1900. The experiment turned out to not be successful, however, and the mill was closed later that year.

By 1905 the town of Dolores was deserted.

Simoni's Store, Cerrillos.

During Cerrillos' early years this building was a hotel.

When I first came to Cerrillos in April, 1970 the first local people I met were the friendly Simoni sisters; Edith and Corrine Simoni, who ran a little store in a room on the bottom floor and lived upstairs. Years later our children always went into their store after getting off the school bus to buy snacks and sodas and play pinball.

I also became friends with their older sister, Emma Simoni Montoya.

Italian names like such as Simoni, Tappero, and Vergolio, along with Spanish last names were the most common in Cerrillos during its early days. I had been told long ago that Italian stone masons were encouraged to move to Cerrillos to build its many quarried sandstone buildings. At the back of the mesa behind the Cerrillos School there are the remains of an old sandstone quarry.

The one-story structure attached to the left side of this building was the Monte Carlos Bar. It had been closed for about seventy years when I made this photograph. Then in 2014, Pat and Kelly Torres bought it. In 2017 Santa Fe County permitted them to open their Black Bird Saloon, a rustic, very popular restaurant serving original recipes made from quality food ingredients, along with beer and wine and other beverages. Pat and Kelly retained the buildings original old wooden interior including the bar.

The coal mining town of Madrid was founded in 1894. A coal fired power plant was built there and so Madrid had regular electric power by 1900, becoming the first town in the region to have domestic electric lights. Its houses were miner's shackles that had been disassembled in Kansas, brought to the site by railroad, then reassembled.

Madrid Gulch coal is geologically associated with the coal that had been mined in Waldo Gulch, the next gulch to the west, since at least the 1830s. They are the oldest coal mines in the Western United States.

All of these local coal deposits were unusual because both bituminous and anthracite coal seams were represented within the coal layers. This is because during the Ortiz Volcano's eruptions 29 million years ago, it baked some of the surrounding Mesa Verde Formation bituminous coal seams into anthracite coal seams.

A bridge was built across the Galisteo River at its confluence with Waldo Gulch in the late 1800s so that a railroad spur could be constructed from the main line, across the Galisteo Arroyo to its south side, and then up Waldo Gulch to the coal mines at Coal Banks.

Then, in 1892, connected tracks were laid up the south side of the Galisteo River from Waldo Gulch to Madrid Gulch, and from there up to the new coal mining camp of Keeseville. Wealthy investors forcibly removed the coal prospectors of Keeseville, threatening to bury alive those who refused to leave their shafts, and Keeseville became the town of Madrid.

The place where these spurs up Waldo Gulch and Madrid Gulch originated from the main railroad tracks eventually became the town of Waldo, located about one and one-half miles west of Cerrillos. The dramatic Cerrillos Hills rock formation, Devil's Throne, stood between the two towns.

A long line of fifteen coke ovens was built at Waldo by Colorado Fuel and Oil Company, along the main railroad tracks there. The coke, made from Madrid bituminous coal, could then be shipped by rail to steel mills. A paint factory was also built at Waldo. This factory utilized zinc mined in the Cerrillos Hills to make white pigment.

By 1906 the Colorado Fuel and Oil Company had folded. But people living in Cerrillos took jobs in Madrid. Madrid and Cerrillos each had a dance hall.

In 1919 a man named Oscar Huber was promoted to Superintendent of Mines for the Albuquerque and Cerrillos Coal Company. Oscar Huber had a different management style than other men in typical American coal mining towns. He created schools, a hospital, a company store, and he provided unlimited free electricity for all of the residents, and many other amenities to assist the local people.

He had a baseball stadium built in Madrid. The Oscar Huber Ballpark was the first electrically lighted ballpark west of the Mississippi in 1922. The local people played for the Madrid Miners, a farm team for the Brooklyn Dodgers. And the Madrid Miners were an inspiration for interest in Minor League Baseball to grow all across the United States. The Dodgers played at the Oscar Huber Ballpark in 1934.

Mary's Bar.

Mary's Bar used to be Tony's Bar. I knew that Tony liked me, though it wasn't obvious. I'd go into his bar when I had a few dollars in order to buy something to eat, usually a bag or two of peanuts. Tony would always kind of gruffly count out my change, leaving me a few cents short. I'd say, "Tony, you owe me another nickel." After some years something happened to him. He got weak. Then when he counted out my change, he'd pour out way too many coins. I would say, "Tony, that's too much," and I'd count the change back to him.

I did not have the opportunity to meet Tony's wife, Catherine Vergolio Tappero, but I have seen her in a nicely framed, black and white photograph owned by Todd and Patricia Brown, in which she is a young woman. In that photograph she is seated on the back of a buckboard wagon, parked on Railroad Avenue with the Cerrillos Railroad Station behind her. That station is no longer there. I was told that Catherine was very beautiful, which is apparent in that photograph. Catherine was my wife's godmother.

Their daughter, Mary, a retired school teacher, and her husband, Leo, moved into the home that is built into this building, behind the bar, from Albuquerque in order to help Tony. Mary had gradually taken over running the bar by the time Tony died.

The priest in Cerrillos at that time conducted Mass daily. Two or three of the town's people would attend on weekdays. Mary walked over to St. Josephs Church to attend Mass every morning.

Leo was rather quiet but very friendly. He had a been part of the invasion at Normandy during World War II. Leo served with a tank battalion. He was captured toward the end of the war and taken to a prison camp in Berlin, but was treated well by the German soldiers who knew by then that they were going to lose the war.

Mary was born at the Palace Hotel. My mother-in-law, Yvonne, told me that when she was in the third grade at the Cerrillos School the school principal gathered everyone in the school gymnasium. His intention was to physically punish three boys. He already had a reputation for being abusive to some of the older boys. When he approached Yvonne's brother, Junior, a seventh grader, in order to beat him, Mary, a young teacher at the time who taught first and second grades, walked up and stood between the boy and the principal.

"Get out of my way. He's done something wrong and deserves to be punished," the principal said.

"No, he hasn't," Mary said, and did not move. Eventually, the principal walked away. He quit his job at the Cerrillos School sometime after that and went to work somewhere else.

As Mary got older her daughter, Cathy, helped her run the bar. I remember seeing Mary walking down First Street in Cerrillos with my father-in-law, Oliver, on a morning when people such as Todd and Patricia Brown were setting up for the annual Cerrillos Fiesta. I had been setting up my table under the Simoni building's porch, preparing to sell my photographs. When Fiesta would begin at noon Cerrillos elder residents would then be honored. Though I don't speak Spanish I am familiar enough with the local dialect to know that Mary was speaking it exquisitely, slowly with Oliver.

Mary died at home on April 20, 2016. On July 19, 2016 Mary would have been 100 years old. Cathy has continued to run Mary's Bar, choosing not to change its name.

Baseball became very popular locally. Cerrillos residents developed their own baseball team, the Cerrillos Blue Jays. Galisteo also had a team. So did Kewa Pueblo and Cochiti Pueblo.

Oscar Huber began the tradition of decorating Madrid and its surrounding hillsides at Christmas. The town's people enthusiastically set up 150,000 electric lights on their homes and upon their creations of large displays in Madrid and upon the hillsides surrounding the town, such as a 90-foot Christmas tree, a large nativity scene, and a gigantic angel blowing a trumpet.

Passing Through Waldo. Standing on the south side of the Galisteo River, looking north over the tops of some of its cottonwood trees growing down in the arroyo, I photographed this freight train as it passed in front of some smaller cottonwood trees, above the arroyo. These trees, standing behind the train tracks, are located at the ghost town known as Waldo. The cliff face of Buffalo Mountain can be seen in the background. Buffalo Mountain was protected from being rubblized into base coarse gravel by a very close decision at the Santa Fe County Courthouse in 2001. That meeting was attended by local activists and a group of elders from Kewa Pueblo.

Airlines altered the course of their planes to allow the passengers to look down upon Madrid from the night skies. The pilots maneuvered around the displays to give passengers detailed views.

Thousands of people came from afar in order to experience what has been called "The First City of Lights." Madrid at Christmas became world famous.

In 1930 artist Paul Lowtz invited his friend Walt Disney to visit Madrid at Christmas. This experience likely provided Walt inspiration for his creation of Disneyland in the 1950s.

During the Great Depression the local people continued to work in Madrid.

On December 7, 1932, an early morning explosion in the Morgan Jones Mine killed fourteen men.

There is an old coal mine opening into a section of Waldo Mesa near its top, above where Waldo Gulch winds around and into the Mesa. I hiked up to this mine one day, just to check it out.

As I stood in the entrance, looking inside, my eyes seemed to adjust suddenly to the darkened shaft, and I was surprised to see how well preserved the old timbers appeared down the descending tunnel. All of the porcelain insulators were still in place near the tops of each of the upright timbers on the right side of the tunnel, which curved downward, and toward the left, into the darkness. It seemed as if I could hear something being drawn into the shaft. I felt eerily drawn to walk inside, but then considered the possibility of hidden rattlesnakes, coiled along the shaft's floor.

Then I noticed that I felt, and had been feeling, air being pulled into the tunnel, like a silent wind, along the sides of my face and my ears. That was part of what was drawing me inside. I wondered why air was going down into an old mineshaft.

Not long after that experience the Bureau of Abandoned Mines arrived in Madrid to inform the community of a plan to fill in many of Madrid's old coal mine openings. I went to the public meeting in Madrid, where I mentioned my experience of feeling air being drawn into a mine opening on Waldo Mesa.

In response the man who was speaking explained that there was a fire still burning deep inside Waldo Mesa from a Madrid mine explosion that had occurred many years earlier, and that because the mines into Waldo Gulch had connections with the Madrid mines, the air that I felt being drawn into that particular shaft was feeding that fire.

I asked if there were any plans to put out this fire.

"No," he said.

Then the man explained to us that in the Eastern United States there are many coal fires. Some of these fires have been burning since the 1700s. Although many attempts have been made to put out underground coal seem fires in the eastern United States, none of these attempts have been successful.

When shafts were buried to starve them of oxygen, the heat from these fires always opened cracks in the rock up to the surface, which then supported their continued ignition. He said that there have actually been fires that have flashed out into hillside infernos in places where the coal seems came to the surface. Afterward, the fires continued their burning underground.

The man also told me about another opening, on the opposite side of Waldo Mesa, that exhaled exhaust from that fire. I knew of the opening he described. It was close to where we lived, almost visible from our house in Waldo Gulch. I hiked up to it one day.

This was a small, rocky opening. No timbers or anything could be seen inside. After sitting there for a short time, the sulfury odor of the fire's exhaust became obvious. I began to feel very comfortable in the warm sunshine, and relaxed. I sat there for a while, smelling the fumes, trying to imagine an image of this underground blaze. Then the thought came to me that I had to get up and walk away. Otherwise, someone was going to find my body there, slumped over in front of that shaft.

Old Madrid Boarding House at Christmas.

After World War II there was no longer a market for the coal mined in Madrid. The railroad stopped their service to Madrid in 1954 and the town population dwindled. In about 1965 the Palace Hotel in Cerrillos was foreclosed. So, before leaving, the Hotel's former owner decided to give away the elegant, antique furniture from all of its rooms to people from Kewa Pueblo. Some of the town's people still remember the many trucks and trailers, loaded with the furniture, driving away down the Waldo Road toward Kewa.

The decision to give away all of that beautiful furniture turned out to be a fortuitous one. In October of 1968 some young people broke into the then vacant Palace Hotel. They built a fire in the fireplace. A townsperson called law enforcement who came and ordered these young people to leave. Apparently, no one thought to put out the fire in the fireplace. During the night that fire somehow spread out onto the wooden floor. The Palace Hotel became engulfed with flame.

The closets full of dance hall girl dresses, the large blood stain on the floor of Dr Palmer's office from a gunshot wound to the head of the famous train robber Thomas "Black Jack" Ketchum, and many other remnants of the past, were consumed, leaving just the building's walls standing. The people of Cerrillos stood outside in the night and watched, crying and helpless.

In 1969 local citizens organized the Turquoise Trail Volunteer Fire Department, which continues to serve and protect the local people.

Today little remains of the town of Dolores. By the late 1930s the remnants of Waldo were being salvaged and it, also, has essentially disappeared. As of the 2010 Census Cerrillos had 321 souls, about the same population as in 1950, including descendants of Spanish and Italian families from its early days. The last Madrid mine closed in 1961. Galisteo is still a small village. It is surrounded by cattle ranches, and the town itself retains a distinctive old Spanish flavor. Its population was 234 in 2010. Many artists have taken up residence there over the years.

Upon the Galisteo Basin's landscape, and within these towns, many famous western motion pictures have been filmed.

When I moved to Cerrillos in 1970 I was told that Madrid, having 35 year-round residents, qualified as a bona fide Ghost Town, based upon the then current population percentage of its original population. I was also told that in 1958 Walt Disney, who was scouting the area for his film, *The Nine Lives of Elfego Baca* (filming began in Cerrillos that year), went over to Madrid. He had become inspired with the idea to make Madrid into a theme park, so he offered Joe Huber, Oscar's son, and current owner of the entire town at that time, $1,000,000 for Madrid. Joe turned Walt down.

Over time the old, abandoned wooden houses began to deteriorate. During the years that I have lived here there have been several tragic fires in Madrid.

Joe began renting houses to hippies. Then, by 1972, he was offering his renters first option to buy the house that they lived in. Many people bought a home and the surrounding property for as little as (as I recall) $1,500. They fixed up and painted many of these old wooden homes.

Eventually the people of Madrid formed a business association and other associations.

Madrid, today, is a unique experience, and a major tourist attraction. The Turquoise Trail passes down the center of Madrid's narrow business district where many old buildings and homes are now art galleries, restaurants, shops, and stores. The Board Walk is open. The old Mine Shaft Tavern is thriving, along with the associated Mining Museum and The Opera House with its concert stage behind the Tavern.

The town has been featured on national television programs. Madrid now has its own radio station, KMRD.

Everything in Madrid is infused with the unique flavor created by the old coal mining town setting, now re-inhabited with the Hippie Migration.

Madrid Church. Until about seventy years ago, this building was St. Anne Catholic Church.

Recently Cerrillos has also begun to see its own traditional kind of old western town revival.

In 1973 Dorothy introduced me to her great great uncle, Santiago N. Mares. He became my best friend.

Every morning Santiago would make a pot of coffee in an aluminum percolator on his old, wood burning cook stove, and fry himself a couple of eggs, sunny side up. Whenever I was there, I would have a cup of coffee with him. I had never drunk coffee, until then.

All of his nephews and nieces would call him Chio. They were always around playing nearby and Chio would have treats for them. Chio had never been married and had no children of his own.

Santiago Mares was born in 1892, on a ranch at the place we now call Bonanza Creek. Much of the large foundation for that original ranch house is still there. The perennial stream, Bonanza Creek, originates in a valley inside some small volcanic mountains, or hills, a short distance to the northwest of the Cerrillos Hills, and are in fact a part of the same eruptive events. The name Los Cerrillos, meaning "The Little Hills," was originally given to these little hills by early Spanish explorers or colonists. The larger portion of the range was named Sierra de Escalante. Much later the title Los Cerrillos, or in English; the Cerrillos Hills, came to refer to that mountain range.

Chio loved to tell me stories about his childhood. As a little boy he worked with his uncles in their wheat fields. At midday the women walked out to them with their lunch. Then Chio and his uncles sat down to eat, and enjoy the respite from their work. And he described to me some of the food the women brought them. He particularly liked a homemade cottage cheese that he and his uncles sweetened with syrup. And there were vegetables from the large garden near their house that the women took care of. After lunch Chio and his uncles would lay down in the field and take long a nap. They awoke around two o'clock in the afternoon and went back to work.

He described to me how the wheat crop was threshed each year. Men from several ranches would get together on a day with some wind and pile the wheat in the center of a large, round corral. Then horses were brought in and the men kept them galloping in circles for hours, around the wheat. Gradually, the vibration from the galloping horses would cause the wheat pile to settle. The lighter chaff would come to the surface and be blown away. Finally, the men went in to store the wheat.

Chio explained to me something about what life was like then: Everyone worked hard from before the sun came up until after the sun went down. Then, everyone would have a late dinner together, and go to bed, and then get up before the sun to start work again.

Santiago's father, Felipe Mares, had, in addition to several hundred head of cattle, well over 2,000 sheep. And Santiago's uncle, his father's brother, had 1,500 or so more. The combined flocks were herded by shepherds in a circuit that took them from their ranch all the way around to circumvent the Cerrillos Hills. Each of these circuits would encompass a year.

Traveling generally east during the spring, along the northeast side of the Cerrillos Hills, they would arrange to be in an area somewhere near where, according to Chio, the Arroyo Coyote crosses the present-day Turquoise Trail National Scenic Byway (New Mexico State Highway 14), at the time in spring when the ewes were about to give birth.

Chio remembered that the shepherds would build a large corral fence in that vicinity with juniper branches. Within this corral the lambs would be born. The rams were separated, kept out to graze all day and all night, while the ewes would graze during the day, then be brought into the corral at night to nurse the lambs.

When the lambs grew strong enough the flock was on the move again. Somewhere past the Cerrillos Hills they circled back, returning down river, toward the west, along the southwest side of the Cerrillos Hills.

Chio told me that by about 1910 so many people had bought property in the area, building fences for their cattle that divided up the land, that his father had to give up his sheep herds. So Felipe Mares opened a meat market in Cerrillos.

When Santiago was a child, his father bought the house in Cerrillos where Dorothy would be raised, and where her mother was born. Felipe bought the house so that his children could attend the Cerrillos School. As a student there, Santiago once received an award for his penmanship, a special pen, which he kept throughout his life.

The family lived in Cerrillos in winter, and at their ranch during the summer.

Chio told me that early in the morning, before going to school, he would drive the miniature buckboard wagon that his father had given him, pulled by two billy goats, into Cerrillos' business area. First, he cleaned two livery stables. Then he went into alleys and side streets looking for bottles that men had drunk from the night before. He put the bottles into his wagon, and then took these bottles to a certain bar where a very gruff bartender was cleaning glasses. The bartender grudgingly paid Chio the deposit for the bottles.

So, before going to school in the morning, the boy, Santiago, earned more money each day than the $10.00 a week that the men earned nearby, working hard labor, taking rock off of Penasco Peak to re-build into the railroad bed.

Chio remembered when the first automobile appeared in Cerrillos. It was purchased by the respected and much liked Dr. Palmer, the town's physician.

One of Chio's memories was of a day when he was young. He remembered walking home from Madrid in the afternoon after working at the Tipple, the large, multistory structure built into the side of a mountain where the coal from the mines was sorted. He began to feel weak as he walked. Someone in a buckboard gave him a ride. Santiago passed out in the buckboard. After that, he remembered his two sisters taking care of him, periodically changing his bed clothes and sheets, which were soaking wet with his sweat. And washing him. That was all that he could remember, until, one day, he awoke and felt better.

It turned out that there had been a typhoid fever epidemic, which Santiago had survived. His two sisters, however, did not survive.

Chio also worked for a while at the paint factory at Waldo. He found his job to be easy and very safe. But he remembered that the company specifically hired men from Kewa Pueblo to perform work that kept them covered in chemical dust throughout the day.

While at work there one day Chio happened to be watching a man placing explosives in a hole that he had drilled into the ground. The explosives detonated prematurely, blowing the man about thirty feet straight up into the air. Several other men ran over to help him, but he was already dead when they arrived.

There was a barber in Cerrillos who was accustomed to cutting the men's hair. But he began to notice that the town's boys had stopped coming in. He observed these boys playing in the streets. Over time, he noticed that they all continued to have excellently cut hair. Finally, he went out and started asking them where they were getting their hair cut. They all had the same answer: "Santiago, he cuts my hair."

So the barber decided to walk over to Santiago's house. When he arrived, he approached Santiago's father and asked him, "Where is Santiago?"

"He's playing with some of his friends over there in that shack."

"Can I talk to him?"

As they approached the little shack the two men saw a line of boys outside leading up to a boy who was sitting down inside the shack, having his hair cut. Santiago was using a pair of scissors to cut the boy's hair, and pieces of broken glass to shave him around the edges. This resulted in an excellent, very professional looking haircut.

Mesa Top Sunset with Elijah. Late one afternoon my son Elijah and I decided to walk up onto a mesa.

60

"I'm next!" the barber said to Santiago, smiling.

"You'll have to wait for all these," Santiago told him in response.

"What are you doing! Get out of there! Go up to the house!" Felipe, not realizing what had been going on, was very angry to see Santiago using broken glass to shave the children. But the barber was impressed.

Santiago had been charging the boys a nickel apiece. Their parents had given them 15 cents to get haircuts. So the children had a dime to spend on whatever they wanted.

The barber asked Felipe if he could train Santiago to be a barber. Felipe agreed, and Santiago became the barber's apprentice. The two cut hair together for a long time. Eventually Santiago opened his own barbershop. The old building with its wooden porch still stands there, in front of the house he lived in. Santiago also cut hair in Madrid.

One day Felipe Mares came home from plowing some vacant fields near the railroad tracks, just upriver from Cerrillos, where he grew corn. Santiago and his brother Tomas saw him coming, and they could see that he was weak and struggling. Felipe asked Santiago to put away the horses, while Tomas helped his father into the house.

When Santiago returned to the house after putting away the horses, his father had died.

Later, he and Tomas followed their father's and the horses' footprints back to the fields. The tracks were very erratic all the way. It was obvious that their father had struggled to get home.

When the coal mines in Madrid began to close and the area lost much of its population, Santiago began cutting hair in Santa Fe. He was very successful, with a loyal clientele. He had many friends in Santa Fe. Everyone called him "Jimmy."

Another story Santiago told me was that as a small child he and some of his family rode their horses out to the mining town of Waldo Viejo, or Old Waldo. He remembered that there were people living there, and there was a store.

But then, he and some family rode out there again, a few years later, and they saw that the town had been abandoned. There was no one there.

The history of Old Waldo is somewhat obscure. There were reports of coal miners at Coal Banks in 1835, the oldest coal mines in the west.

My old friend, the artist Jerry West, grew up in the upper Gallinas Arroyo area in the north central portion of the Galisteo Basin. This was where his father and uncle had homesteaded ranch land. Jerry once told me that he remembered stories from those days about coal mining that was already taking place at Coal Banks by the 1830s. Jerry told me that the area was called Coal Banks because the coal seams are visible where they emerge to the surface, on the sides of the mesas that define what is now named Waldo Gulch and, also, Miller Gulch, the next drainage to the west.

Jerry also commented that the Civil War Battle of Glorieta Pass, which was fought March 26-28, 1862, near the site of the ancient Pueblo Cicuye, had been fought in part because of the wealth of the area, and that the coal resources at Coal Banks were a significant aspect of this wealth because of the extreme scarcity of that industrially important product in the Western United States. The Battle of Glorieta Pass has been dubbed "The Gettysburg of the West."

I knew of the place. Todd, Eric, Rick and I used to walk out there to see the old ghost town all made of sandstone, and the old mine openings, an area of bricks that I thought were remains of a single coke oven, and the evidence of what had once been a railroad bed. There was no road to the place at that time.

Then, in 1977, a man named Bob Taylor bought 150 acres that included most of the ghost town. He hired Henry Trigg to bulldoze a road into it. There was already an old dirt tank there, in Waldo Gulch, which Henry had previously built with his backhoe, in order to water his cattle.

Our friend, Sonny Pasero, was helping with the restoration of one of the ruins, using the original technique of placing slightly moistened mud between the locally quarried sandstone. He invited Dorothy and me to take jobs on this construction. It was summer. We got up early, and quit around 2:30 in the afternoon, when it started to get hot.

Bonanza Creek Cottonwoods. El Rael de Los Cerrillos was the first documented mining district in the American west and was located at Bonanza Creek. Santiago N. Mares, a descendant of early Spanish immigrants to northern New Mexico, was born at Bonanza Creek and raised there as a child.

Later in the year when the little house, which turned out to have once been the store that Santiago remembered, was nearly finished, Bob Taylor invited Dorothy and me and our children to come and live there as caretakers. Bob had wanted to have animals there. Dorothy and I each had a horse, and we had several goats. I believe, also, that Bob liked us, and trusted us.

We reassembled a small, old log cabin that Bob had found up on Rowe Mesa, which was to be our tack room. And we built a couple of large corrals with simple, wooden stalls. Bob bought us a water tank for the horses to drink from.

Next to the corrals we built a goat pen. It had a small sandstone ruin inside that we made into a goat house. And, directly across the dirt road from the goat pen, the road that led up to the rebuilt structure that was to be our home, we built a chicken coop with a fence around it, and the top covered with chicken wire to keep out the coyotes.

We lived there from January, 1978 to September, 1989.

The house had three rooms that included a small bathroom in the back with an old pot-bellied stove, a sink, and an old-fashioned claw foot bath tub. The running water, and all of our water, had to be hauled in, as there was no well. There was a water tank positioned uphill behind the house.

The bedroom was in the middle. It had a beautiful stone fireplace that Eric built. One of our children, who came too suddenly for us to drive to the hospital, would be born in front of that fireplace, just as the rising sun shown in through our bedroom's east window.

In the kitchen there was an antique wood cook stove, a propane refrigerator, a sink, and an antique table and chairs. The house was lit by kerosene lamps. There was no electricity.

There was an outhouse down near the corrals.

It was remote. Once we took the boys backpacking into the Pecos Wilderness for a week. One day, after returning home, Dorothy and I were sitting on the front porch. "You know," she commented, "There are way more people in the Pecos Wilderness than there are around here."

Not long after moving there a long-time friend, Susan Silver, and I went out on horseback one day to herd back some horses that were running loose. These horses had belonged to several people we knew who had intentionally let them go. Susan was an excellent rider. She had worked as a stunt woman on western motion pictures. We managed to locate the horses after a couple of hours. They had been living in the wild for several years. We began to maneuver around them in order to begin herding the horses back to the ranch.

Just when we thought they were going along with us, they turned back, suddenly, each horse heading in a different direction. They did this sort of thing again and again, outsmarting us.

But, thanks to Susan's riding skills, and her uncanny ability to anticipate the horses' strategies, we eventually got them all the way across the valley, up onto the mesa, across the mesa top, then down into Miller Gulch, down Miller Gulch a ways, and over into Waldo Gulch and the little ranch.

I ended up keeping two mares. The next year both mares had colts.

I let our small herd of horses roam on the many thousands of acres of open pasture, among the mesas, canyons, and valleys behind Waldo Gulch, for about a month or so, once or twice each summer. I would go out on foot when I wanted to check on them or bring them home. I'd find them on a mesa top, or down in a valley. I always brought a bridle and typically got on Dorothy's horse, Rebel, because he liked to run, and I'd bring the horses running back.

Then, one time, I went out and looked for hours, but I could not find them. Finally, I dejectedly headed home.

When I got home, there they were! After being gone for over a month I found them waiting for me next to the corrals. I was amazed.

Chimayo Youth Corps. In 2005 and again the following year an arrangement was made to have young people who were part of the Chimayo Youth Corps come to Cerrillos to help create trails in the Cerrillos Hills Park.

Road to the Cerrillos Hills Park, Sunrise, from Penasco Peak. The road to the Cerrillos Hills Park is at the far-left side of this photograph, rising up out of the San Marcos Arroyo and into the Cerrillos Hills. Grand Central Mountain, and Cerro de la Cosina to its right, can be seen with the rising sun shining upon them in the background. The Cerrillos Hills Park Coalition, along with all other local citizens organizations and a majority of Cerrillos residents, are currently, as of this writing, working to prevent a proposed cell phone tower on Penasco Peak. Two previous cell phone tower proposals have already been defeated.

I thought, "They knew I wanted them to come home, so they made the trip on their own." It was just too much of a coincidence. Well, I wasn't sure whether it was a coincidence or not.

But over the course of the years, that same thing happened again and again, quite a few times, though they never came home on their own when I wasn't out looking for them. Most times, when I went looking for them, I could find the horses. But there were other times that I looked and looked and could not find them. Then, when I got home, there they would be, waiting at the corrals, after having been gone for a month, sometimes almost two months.

They could read my mind. Or they had an empathy with my being.

During the last twenty-five years or so, as some of the cattle ranches within the Galisteo Basin have disappeared, its population has, finally, once again reached an order of magnitude similar to the basin's Classic Pueblo Period. Many of these people have taken an interest in preserving the basin's beauty and resources from inappropriate development and poor planning.

Most of the prehistoric and historic mining that has taken place in the Galisteo Basin occurred prior to modern mining methods. Many of the old mines are preserved in the Cerrillos Hills State Park and the Ortiz Mountain Preserve as mementos of the past.

The one exception is the Gold Fields Mine.

Between 1980 and 1987 Gold Fields of South Africa Ltd. extracted between 231,000 and 250,000 troy ounces of gold from a volcanic vent in the Ortiz Mountains. The gold was extracted using a cyanide-leach process. A mountain named Cunningham Hill was destroyed. It is now a pit 385 feet deep, partially filled with polluted water requiring continuing remediation.

Ross Lockridge and Ann Murray met when they were students at the San Francisco Art Institute. They moved to Cerrillos in April of 1972 and began volunteering in the community.

While supporting themselves as artists their activism evolved, and the timing of this proved to be serendipitous. Others of us have learned to depend upon them for their many efforts to stay abreast of new developments that might impact the Galisteo Basin.

From 1974 to 1978 they, along with historian Marc Simmons and Deidera Hazelrigg, initiated efforts that saved the Cerrillos Hills from a proposal for in-situ copper leach mining. They (and we the community) protested and prevented the Cerrillos Hills from being rubbled with high explosives in order to leach out copper with sulfuric acid. Ross and Ann helped incorporate the Concerned Citizens of Cerrillos (CCC) in 1976, during this struggle, and were active with the CCC in drawing attention to and resisting the adverse environmental effects of the heap leach gold mining in the Ortiz Mountains from 1978 to 1988.

Thereafter Ross and Ann and the CCC worked with Friends of Santa Fe County and the New Mexico Environmental Law Center toward reclamation of the Gold Fields mine site. The Ortiz Mountain Educational Preserve was created during this time. The couple helped stop a proposed strip mine of coal in Miller Gulch west of Waldo Gulch, and another attempt at heap leach mining in the Cerrillos Hills. And they helped prevent the return of another heap leach gold mine to the Ortiz.

Gravel mining became a looming issue in the Galisteo Basin, and Ross and Ann with others of us in the community formed the Rural Conservation Alliance (RCA) to stop a large and poorly sited hard-rock crushing operation in the Cerrillos Hills, which would have greatly impacted the Hills, the village of Cerrillos, and the Turquoise Trail.

They and the RCA then helped lead the struggle to save Buffalo Mountain, a colorful and dramatic landmark in the Cerrillos Hills, from a proposal to remove it by crushing the mountain into base coarse gravels for highways and construction materials.

The Village of Los Cerrillos and the Ortiz Mountains from the Cerrillos Hills Park. Some of the village can be seen in the left middle ground of this photograph if one looks closely among the trees. A little further to the left, the old gymnasium and remnants of the quarried sandstone Cerrillos School, located across the Galisteo River and the Turquoise Trail National Scenic Byway from the main part of the village, can be seen inside the school's large, square wall. The deep snow seen here is from the record snowfall mentioned in the caption for St. Josephs Church in Snowstorm photograph. This photograph was taken early on the morning after the snowstorm ended: January 1, 2007.

That proposal was narrowly defeated in 2001 with a large community turnout that overflowed the chamber of the Santa Fe County Courthouse, filling the Courthouse lobby as well.

A group of elders from Kewa Pueblo surprised us by walking in during that meeting. They encircled the interior of the courtroom chamber in single file, and then sat down at the front of the audience, in a row of chairs that had been reserved for them. Their former governor then stood up and spoke on behalf of saving Buffalo Mountain.

In response to another gravel mining proposal Ross and Ann helped the RCA in a successful nomination of La Bajada Mesa and Escarpment to be included in the New Mexico Heritage Preservation Alliance 2003 List of Most Endangered Places. More recently they have worked with an enormous community response that has stopped a proposal for large scale exploration and drilling for oil and natural gas in the Galisteo Basin.

The Concerned Citizens of Cerrillos, a 501(c)(3) non-profit, became an umbrella organization for Drilling Santa Fe, a non-profit started by a couple originally from Texas named Johnny Micou and Nancy Seawald, specifically to keep an eye on the oil and gas industry in New Mexico. This monumental, grass roots effort resulted in Santa Fe County establishing the strictest laws at the time governing oil and natural gas exploration and extraction in the United States, and the strengthening of state law in New Mexico, which is a major oil and natural gas exporter. Those of us who live here know that without Ross and Ann's leadership, and the efforts of many like-minded residents, the Galisteo Basin would be a drastically different place today. They have worked with community organizations and Santa Fe County to develop community plans, and projects and activities that benefit the citizens and community. In 2001 Ross was recipient of the Griff Salisbury Environmental Protection Award for his work chairing a citizen's advisory committee that brought about context sensitive designs for the reconstruction of the Turquoise Trail National Scenic Byway.

Ann envisioned a park in the Cerrillos Hills. And so, she wrote a proposal to protect the Cerrillos Hills as a park. During community planning in 1997, thanks to Ann's vision, she and Ross joined with local citizens to form what became the Cerrillos Hills Park Coalition. Thereafter the Coalition began meeting together in order to create a proposal for a park in the Cerrillos Hills.

After Santa Fe County passed an Open Space Bond in 1998, they accepted this proposal, and in 2003 the Cerrillos Hills Historic Park was officially opened. There were many who spoke at this ceremony celebrating the opening of the park.

The guest speaker was Stewart Udall, former Secretary of the Interior during the Administration of President John F. Kennedy, and U.S. Congressman. Stewart provided us an inspiring message about our park.

Open space was also created around Petroglyph Hill. Among the charter members of the Cerrillos Hills Park Coalition, founders of the Park, were Ross and Anne, and a man named William "Bill" Baxter. Bill was a local historian, author of *The Gold of the Ortiz Mountains, A Story of New Mexico, The West's First Major Gold Rush*, and *Gold and the Ortiz Mine Grant, A New Mexico History and Reference Guide*.

Everything that Bill describes in his books is historically authentic. The Gold of the Ortiz Mountains has a focus upon the gold mining, and the town of Dolores. Yet, many other facts are described. For instance, in *The Gold of the Ortiz Mountains*, Bill states, "at least up to 1846 the Ortiz Mountains were host to a healthy population of grizzly bears. The last grizzly bear in New Mexico was killed in 1931."

Ross and Ann in the Cerrillos Hills Park. The Cerrillos Hills Park was originally the idea of Anne Murray. Over many years Ross and Anne have inspired people to work with them toward many accomplishments that have protected the Galisteo Basin from attempts to destroy its natural beauty. These efforts included the creation of the Cerrillos Hills Historic Park, which then became the Cerrillos Hills State Park.

Bill Baxter was the single most significant contributor to my research for this book. Among his many involvements within the Galisteo Basin, including the Ortiz Mountains Educational Preserve, Bill has, without a doubt, done more to shepherd the creation and development of the Cerrillos Hills Park than anyone. This includes negotiating with Santa Fe County and the Bureau of Land Management; the designing, building and maintaining of the Park's trails; researching, writing and displaying on interpretative signs and brochures detailing the Park's historic, prehistoric and geologic significance; and giving walking history tours to many sites.

When New Mexico State Parks began to express interest in creating a Galisteo Basin State Park, citizens of the Upper Basin were cool to the idea of welcoming the public because of their concerns about the sensitive nature of sites, such as the Open Space around Petroglyph Hill, and some of the ancient pueblo ruins in the area. But Cerrillos Hills Park Coalition members and people in the Cerrillos and Madrid area were more receptive to including the Cerrillos Hills Historic Park into this proposal. It was already open to the public. Bill was especially enthusiastic about this possibility.

Bill had long envisioned the Galisteo Basin becoming a National Monument, and he saw State Park status as a step in that direction. The Cerrillos Hills Historic Park became the Cerrillos Hills State Park in September, 2009, and the Park Coalition developed a rewarding relationship with Park Rangers Sarah Wood and Peter Lipscomb of New Mexico State Parks. So, Ann's vision of a park in the Cerrillos Hills ultimately led to the creation of the Cerrillos Hills State Park.

Today the Cerrillos Hills Park Coalition continues to partner with Santa Fe County, helping the county work through complications as it attempts to acquire the property that would include and protect Mount Chalchiuitl.

Working closely with Bill, and on behalf of the Cerrillos Hills Park and the Ortiz Mountain Educational Preserve, especially by helping create, manage and maintain their minimal physical infrastructure and protecting their natural integrity, has been Todd Brown. Todd arrived in Cerrillos, as a seventeen-year-old from Long Island, around the same time I did, in the spring of 1970.

He and his wife Patricia, who is also a friend of mine from those early days here, own Casa Grande Trading Post, the Cerrillos Mining Museum, and the Cerrillos Petting Zoo, all incorporated with their home into a three story, pueblo style structure that they built out of adobe bricks, many of which were created on site by forming the mud and straw in a traditional form, then standing the bricks in the sun to dry. They have a legal turquoise mining claim in the Cerrillos Hills. Todd makes jewelry from this turquoise, as does his son, Andy, which Todd and Patricia sell in their store. Todd learned to make jewelry many years ago, from Leo.

In 1993 a gifted young man moved to the area. His name is Jan-Willem Jansens, a native of The Netherlands. He is a Professional Landscape Planner with a specialization in ecological restoration of degraded lands.

He developed an interest in the Galisteo Basin in 1997, and began performing ecological restoration projects and conservation planning there. He spearheaded many different kinds of ventures. In 1998 this work became known as the Galisteo Watershed Restoration Project, funded by the McCune Charitable Foundation.

Cerrillos Hills and the Arroyo, sunset. Grand Central Mountain, with Cerro de la Cosina to its right side and Cerro Bonanza to its left, have the light of the setting sun upon them. Before them and to the left in this photograph the steeple of St. Joseph Church and trees in the village of Cerrillos can be seen. This red sunset light on the hills is akin to the light that inspired the early Spanish explorer Antonio Valverde y Cosio to name the Sangre de Cristo, or Blood of Christ, Mountains.

A large number of local people were inspired to become involved with this project, which received grants from the State of New Mexico twice as well as a grant from the US Fish and Wildlife Service.

Also in 1998, Jan-Willem began contributing as a contractor to Earth Works Institute, a New Mexico non-profit organization that had been established in 1984 to provide community education, and to promote and demonstrate strategies for sustainable living, land restoration, and community stewardship. Jan-Willem became this organization's Executive Director in 2004, directing its efforts toward the formation of the Galisteo Watershed Partnership, which, after a great deal of effort, was founded in 2005 with local citizens, land owners, ranchers, and environmental organizations. The Partnership was officially endorsed by Santa Fe County in 2005.

In 2005 the New Mexico State Legislature allocated $50,000 toward a Conservation Plan for the Galisteo Basin.

In 2008 Earth Works Institute and the Galisteo Watershed Partnership organized and conducted several meetings about wildlife conservation in the Galisteo Basin area, and the importance of connectivity across the landscape for the needs of free roaming wildlife. In 2009 these efforts led to the establishment of the Galisteo Wildway.

The Galisteo Wildway, Jan-Willem explained to me, is a new segment in a network of wildlife conservation areas that make up the Western Wildway, which is stewarded by members of the Western Wildway Network, an international coalition of wildlife conservation groups that coordinates wildlife linkages between protected hub areas from the Alaskan and Canadian North Coast, down through Western North America, all the way to the Jaguar Preserve in the Sonoran Desert of Mexico.

Jan-Willem also became very involved in the community response to prevent large scale oil and gas exploration in the Galisteo Basin, that I mentioned above.

And through the Galisteo Watershed Partnership, Jan-Willem influenced the State of New Mexico regarding its recent selection for the new Railrunner commuter train route from Albuquerque to Santa Fe; to choose to build this rail line up the center of Interstate 25, rather than accepting another proposal to build the rail line from Cerrillos north along the Turquoise Trail National Scenic Byway (New Mexico State Highway 14). This would have destroyed portions of the Galisteo Sandstone Formation, along with other pieces of the natural landscape.

Between 2008 and 2010 Jan-Willem was one of the authors of the Galisteo Watershed Conservation Initiative. This document identified and mapped key conservation values for the Galisteo Basin, and it influenced Santa Fe County's way of describing and managing lands in its Sustainable Growth Management Plan.

Recently Jan-Willem explained to me that approximately 130,000 of the Basin's 467,000 acres is now under some form of protected status or responsibly managed, nearly 28% of the Galisteo Basin.

He has also taught me something that I mentioned earlier in this book: the Galisteo Basin is actually the convergence of four different ecoregions. It therefore has a great diversity of plant life, and includes wildlife corridors for many different kinds of animal and bird species to migrate from place to place.

Todd Brown's Cerrillos Turquoise and Jewelry. Todd mined the Cerrillos Hills turquoise in this display, which is located inside his Casa Grande Trading Post, and he created the silver and turquoise jewelry.

In April of 2016 The Bureau of Land Management purchased 365 acres of Burnt Corn Pueblo from a private land owner. The $1.5 million came from the federal Land and Water Conservation Fund. The transaction was accomplished with help from the Trust for Public Lands, a congressional delegation, and the public. This acquisition placed the entire site of Burnt Corn Pueblo onto public land, and opened up access to more than two thousand other acres of public land in that area. This included Santa Fe County Open Space, BLM, and state land. The acquisition also opened access to Galisteo Basin Preserve land.

As reported in *The Santa Fe New Mexican*, Senator Martin Heinrich said, "This incredible success wouldn't be possible without the advocacy of pueblo leaders and community members who've worked for years to have this site protected."

Many years ago, Bill Baxter came to visit me. Knowing my love of photography, he wanted us to watch a projection of some of his slides together, a slideshow. We sat on the floor while he arranged his slides and his little projector, and he explained to me the subject of his presentation, which included aspects of his life that I had not previously known. When Bill was a young man, he became a member of the Peace Corps. For that service he went to Ethiopia where he taught school. When he finished that service he traveled to Egypt, and from there he began a walk, a walk that would not be possible now, at the time of this writing. From Egypt he walked into Israel, and from there, across the Middle East and into India. He liked India and stayed there for a while before returning home.

The slide show he wanted me to watch with him was of this walking journey. There were expansive images of large mosques, from both the inside and outside, dramatic landscapes and seascapes, the great Buddhas carved into the Bamyan Cliffs which have since been destroyed, and his portraits of individuals and groups of local people in their traditional clothing, including many Pashtun people; I especially remember two old Pashtun men, one from Afghanistan, one from Pakistan.

What makes a place sacred? What makes a land a Holy Land?

Back in 1970 when I told people that Jesus led me to this place, to my surprise, for reasons other than mine, the most common response was, "I was led here too." Or, "I've always felt like I was sent here." A girl told me that the founders of the Lama Foundation chose their site near Questa, New Mexico because they believed that Northern New Mexico was a Holy Land. Many of these friends are still here and still explain that they were led or brought to this place.

Perhaps this place was sacred to the Paleo Indians who walked into the Basin to hunt the megafauna and giant bison; the Archaic Era people who made annual migrations among its mountains, mesas, hills, and streams; the Anasazi People who drew petroglyphs on the volcanic rocks; the Franciscan Friars who gave their lives here for their faith; the Spanish settlers who traveled to this place by choice, or their children who were born here and their descendants who live here still.

What constitutes a Holy Land? Does it require that people of different ethnicities, languages, and religions commit acts of injustice, persecution, vengeance, and warfare while attempting to control the land? The Galisteo Basin has a history of these very acts, and possibly a prehistory as well, between rival tribes.

When they returned to New Mexico after the Pueblo Revolts, the Spanish became careful to no longer interfere with the religious practices of the Native Americans. In 1821 Spanish Governor Melgares announced that the "minority" of the Native Americans was ended; they would now be regarded as Spaniards in all things. Perhaps the answer to the question, what constitutes a Holy Land, lies with each one of us.

Today the Galisteo Basin is a place where people from different backgrounds can demonstrate their appreciation for this earth, which has provided us all a home, by working together to protect the landscape along with its water and ecosystems, as well as the monuments, artifacts, and remains of its ancestral people.

I believe that the entire earth is sacred, and very fragile. For me, the Galisteo Basin is the palm of the earth's hand, which has sheltered and nourished me, and healed me. I love this place. My wife Dorothy, who has lived here her whole life, and I have been able to raise our children within its heavenly light, safely and in peace.

Dedicated volunteers. I took this photograph in 2000 when a large group of us returned to Cerrillos after a hard day of trail work in the Cerrillos Hills. Among the many volunteers here are Ross, left of center, Anne, right of center, and Bill Baxter, off to the left. The current owner of the land upon which everyone is standing, and which borders the park at the end of the Elkins Trail, has, in the year 2021, made an offer to donate her property to the Cerrillos Hills State Park.

In 1972, before I ever had a horse, I decided to fast and explore the Upper Galisteo Basin. It was May, and it was hot and dry that year. Four dogs went with me: Norton (originally named Hassan), Piggy (originally named Little Sister), God's Way, and Red Sky. We went first up into the Cerrillos Hills, to Eric's old underground house, so I could retrieve my sleeping bag. Big Joe Merrick happened to be inside. We talked for a while. I told him what I was planning to do. Then the dogs and I headed east, stopping at Bart Durham's ranch.

"You should go see the Bear Petroglyph on top of Petroglyph Hill," Bart told me.

"How do I get to Petroglyph Hill?"

"Just keep going up and when you can't go any higher, you'll see the bear."

The dogs and I spent the first night near the old Sweet Ranch, which was abandoned at the time, at the Petrified Forest. The next day we continued east, moving up, until we got to the top of the highest hill. And there it was, the Bear Petroglyph, the most incredible petroglyph that I had ever seen.

Bear Petroglyph. Though not a prehistoric Native American petroglyph, the experience of seeing the Bear Petroglyph on top of Petroglyph Hill is hard to put into words. Many years ago, someone told me that the Bear Petroglyph was created by a shepherd during the 17th or 18th century. In their book, Burnt Corn Pueblo, Conflict and Conflagration in the Galisteo Basin, *James Snead and Mark Allen included a photograph of the Bear Petroglyph with the caption, "Figure 714, Historic petroglyph of a bear, signed by 'JSA.'"*

Descending the south side of Petroglyph Hill, with Cerro Pelon standing across the Galisteo River from us, we found a valley with the remains of an old homestead and a windmill. Then we headed for the Galisteo River (Creek). We crossed the river. I went up to the old Silver Ranch house and knocked on the door. Annie was there. (Thirty-six years later, Annie taught our daughter to ride a horse.)

I asked her and her boyfriend for some dog food, which they gave to me. I spent the night in a small cave in an outcrop of sandstone near the Finger Lakes. It was just large enough for me. The dogs slept nearby outside.

In the morning we continued south, up the east side of the Arroyo Chorro. From the top of a rocky hill that stood in the Chorro I could see the Cash Ranch, Hazel's house down in the distance. I didn't realize then that two years later Hazel Cash would shoot me for trespassing upstream from the San Lazaro Pueblo ruins.

We left the hill going southeast. I wanted to summit Cerro Pelon. After crossing a small drainage, the dogs and I began walking uphill. I carried my rolled up sleeping bag over my shoulders, my head looking down at the ground. I wasn't paying much attention to where I was going.

I noticed some pottery sherds. They gradually got bigger as we moved up the hill. They looked new, like they had been recently glazed and fired, a lot of red clay, none of the black on white shards I had often found.

All of a sudden, I became startled by how large the sherds were getting. Many were nearly half the size of a pot. I stopped walking and looked around.

We were standing near the hilltop. It had a white sandstone cap just above and in front of us, a little to our right. In the base of the sandstone outcrop was a natural cave. The pottery sherds were scattered before its entrance and on down the hill. The cave was big enough to sleep several people. At its back was a natural chimney in the rock, with a large deposit of ashes beneath.

Sticking out of the ashes there was the upside-down rib cage of a medium sized animal, probably a deer. I walked up to the cave. I touched my thumb and forefinger to the tip of one of the ribs, to gently test its flexibility, trying to determine its age. The end of the bone unexpectedly snapped off with almost no effort, surprising me. I knew that, being so brittle, it must have been very old.

I thought to myself then that my discovery epitomized for me the meaning of the term "Land of Enchantment," as it still does even now.

The dogs and I continued climbing along the slope of Cerro Pelon's north edge, through the heat of the day, reaching the top late in the afternoon. I was very thirsty so it was a great relief to see the large stock tank near the bottom of the cliffs before us. I took in the view, examining the landscape in all directions, resting for a while. Looking straight ahead, toward the Basin's Eastern Mesa Highlands, the afternoon sun lit everything beautifully for me.

Then we started our descent. I had to help Red Sky, the smallest dog, over the large boulders and down the rock faces.

I envied the dogs, and marveled. They stood inside the stock tank and continuously lapped up the water for twenty minutes or more, while I was filled up in twenty seconds or less, and still very thirsty. I hadn't brought a canteen, or anything except the sleeping bag. We hung out until I couldn't drink anymore.

In the long, north south valley in front of Cerro Pelon, south of the village of Galisteo, the Galisteo Basin has a confluence of its major tributaries. One tributary, Galisteo Creek, heads northeast. It eventually enters the Sangre de Cristo Mountains where it drains Thompson Peak and other peaks. I picked the other tributary, San Cristobal Creek. That one continues straight upriver, east, across this valley of the upper Basin and into the Eastern Mesa Highlands.

We walked on, crossed Highway 41, stopping for the night a way farther, above the San Cristobal arroyo on its north side, when the western horizon had just a band of red after sunset.

I awoke the next morning and went down into the arroyo to get a drink of water. Then we hiked upstream.

The sky was cloudless and blue, like every other day of my journey. I remember the last little pool of water, mixed with cow pee, as the day started to warm. As we reached the mesas, I saw that the arroyo in the savanna grasslands became a canyon, among the other canyons within the Galisteo Basin's Eastern Mesa Highlands.

We entered the canyon, which gradually began ascending. After hiking for some time, the canyon veered toward my left, and then I saw a rock overhang in the left canyon wall. I climbed up into it.

There were some large obsidian flakes, and what looked like a small, green scraping tool with a round, sharpened notch on the inside. Further along, up the canyon, all the big piñon trees were standing and dead; just the junipers were alive and healthy. The canyon started to flatten out.

I wanted to reach the canyon's source. But, maybe around two in the afternoon, I had to turn back because of thirst.

I couldn't wait to get back to that little piss filled water hole. I moved faster and faster. Finally, there it was. I took a long drink.

By the following day the dogs and I were back at the Finger Lakes. I was so weak from dehydration that I spent the entire next day there, every so often going to drink, counting my swallows so that I would drink more than I might otherwise. Sometimes I would go over to the second Finger Lake which had a very old, wooden dock. I'd lay on the dock and wait for the big catfish to swim by.

The next day, my seventh day out, I started to feel a little stronger. So midmorning I broke away from my measured drinking and went looking for the cave in the hilltop. The dogs and I reached the top of the hill that I was sure was the one, and it had a white sandstone cap, but there was no cave. I couldn't believe it. We ended up finding lots of hills with sandstone caps and each time I was sure that it was the one.

I never saw the cave again. That night there was a full moon. I lay in the little cave by the Finger Lakes and couldn't sleep.

Suddenly, in the middle of the night, I realized that Piggy had disappeared. Then I recalled that her recently weaned litter of puppies were back at home.

So, I got up, rolled up my sleeping bag, and Norton and God's Way and Red Sky and I walked home together in the moonlight.

Interior of Monte Carlos Bar. This black and white photograph was taken during the 1930s or possibly 1920s. People in this photograph include, from left, someone named Baker in a dark suit, and then the short man behind the bar is Gene Montoya the owner of the bar, and then Juan Ramirez, standing behind the bar, whose daughter, Vangie Martinez, still lives in Cerrillos. Tony Simoni is at center with a hat on, and Emma Simoni Montoya is behind the bar. She is the daughter of Tony Simoni and wife of Gene Montoya. Then Tomas Mares, great grandfather of my wife Dorothy, with a hat on and looking toward the camera. Farthest to the right is a man named Cedillo. Photograph courtesy of Yvonne Perea.

La Bajada Mesa and Escarpment, from La Bajada Mesa. Looking across the escarpment toward the west, one sees its southern portion which drops off with vertical cliffs of basalt toward Galisteo Creek, nearby to the left. In the middle background stands Mount Titilla, La Bajada's most prominent and well-known volcanic vent.

Many years ago, I asked my father-in-law, Oliver, "What is the English translation of Titilla?"

"Tit!" was his dramatic reply, indicating to me that the answer to my question should be quite obvious.

PUEBLOS OF TODAY IN AND NEAR THE RIO GRANDE VALLEY:

Tanoan Pueblos of today consist of three language branches. The Tewa Tanoan speaking Pueblos are Tesuque, Nambe, Pojaque, San Ildefonso, Santa Clara, and Ohkay Owingeh, all located in the Rio Grande and tributary drainages north of Santa Fe; the Tiwa Tanoan speaking Pueblos are Picuris and Taos located north of Ohkay Owingeh, and Sandia and Isleta Pueblos located south along the Rio Grande, just north and south of Albuquerque respectively. The now abandoned Piro sites, located further to the south were also Tiwa; the Towa Tanoan language branch is spoken by the people of Jemez Pueblo located in the upper Jemez River drainage.

The Keresian speaking people are related in modern times to the three central Rio Grande Pueblos of Cochiti, Kewa, and San Felipe, all south of Santa Fe, and to Santa Ana Pueblo which is located on the Jemez River drainage near to the Rio Grande and the town of Bernalillo, and to Zia Pueblo, located further up the Jemez River, between Santa Ana and Jemez Pueblos; and to Laguna Pueblo located on the Rio Puerco River drainage, and Acoma Pueblo.

www.ingramcontent.com/pod-product-compliance
Lightning Source LLC
Chambersburg PA
CBHW041300210326

41599CB00007B/254